Fires in Conventional and Electrified Vehicles

Theory, Prevention, and Analysis

Fires in Conventional and Electrified Vehicles

Theory, Prevention, and Analysis

Erbis Llobet Biscarri

SAE INTERNATIONAL®

400 Commonwealth Drive
Warrendale, PA 15096-0001 USA
E-mail: CustomerService@sae.org
Phone: 877-606-7323 (inside USA and Canada)
              724-776-4970 (outside USA)
Fax: 724-776-0790

Library of Congress Catalog Number 2024941566
http://dx.doi.org/10.4271/9781468607956

Information contained in this work has been obtained by SAE International from sources believed to be reliable. However, neither SAE International nor its authors guarantee the accuracy or completeness of any information published herein and neither SAE International nor its authors shall be responsible for any errors, omissions, or damages arising out of use of this information. This work is published with the understanding that SAE International and its authors are supplying information but are not attempting to render engineering or other professional services. If such services are required, the assistance of an appropriate professional should be sought.

ISBN-Print   978-1-4686-0794-9
ISBN-PDF   978-1-4686-0795-6
ISBN-epub   978-1-4686-0796-3

To purchase bulk quantities, please contact: SAE Customer Service

E-mail: CustomerService@sae.org
Phone: 877-606-7323 (inside USA and Canada)
              724-776-4970 (outside USA)
Fax: 724-776-0790

Visit the SAE International Bookstore at books.sae.org

Publisher
Sherry Dickinson Nigam

Product Manager
Amanda Zeidan

Production and Manufacturing Associate
Michelle Silberman

Dedicated to Celi, my lifelong love, friend, companion, and spouse. I express profound gratitude for her encouragement, unwavering support, valuable suggestions, dear corrections, and boundless fortitude.

Contents

Section 3: Analysis

Chapter 12 – Recommended Analysis Method

Chapter 13 - Data Collection and Analysis

Foreword

Since the invention of the automobile, the risk and the cases of fire exist due to the use of flammable liquids and later due to the increase in the number of electrical and electronic devices. At the same time, the car manufacturers worked on the analysis and understanding of these events that jeopardize the safety of users in order to better prevent this risk as much as possible.

Mr. Erbis Llobet Biscarri, with his experience in this field both at automotive equipment manufacturers and then at a major car manufacturer, decided to establish a guide dealing with the physical phenomena leading to the fire of a vehicle, with the methods of analysis to identify the root causes and with the principles of prevention of these risks, both for vehicles with traditional combustion engines and the latest propulsion technologies (hybrid, electric, etc.).

Having had the pleasure of working for many years with Mr. Biscarri whose professionalism I appreciated and as an expert in the field of vehicle fires, I can only rejoice and congratulate Erbis for having conducted this thorough work of synthesis of the topic of vehicle fires.

I am convinced that the reader will find in this book all the elements of understanding and knowledge of the event of vehicle fires.

This book will be a valuable help for experts in vehicle fire analysis as for the design and validation teams of new vehicles from car manufacturers.

Aubert George, Expert in vehicle fire analysis and prevention, France.

Acknowledgments

Acknowledging every individual who has influenced my professional journey and contributed to this book would be insurmountable. I want to begin by expressing my gratitude to those who may not be explicitly mentioned here solely due to memory limitations. With this clarification, I extend my sincere appreciation to my parents, who have been the foundation of it all and have played pivotal roles in shaping my path. Additionally, my heartfelt thanks go to the myriad of influencers, including professors, experts, mentors, colleagues, subordinates, and students. Embracing the truth in the adage that teaching is the best form of learning, each class I instruct and every analysis report I undertake provides me with an invaluable learning experience, revealing new dimensions, unexpected insights, instigating details, and diverse perspectives.

A special note of appreciation goes out to the former Peugeot-Citroën *plateau* ECA/CCA members and contributors: Aubert George, Benoit Corbic, Agnès Lorne, Sandro Diniz, Rinaldo Tavares, and Leanderson Nascimento, among others. Being part of this exceptional team was an honor I take pride in. The valuable understandings gained from our interactions and collaborative efforts resonate throughout this book.

I am grateful for the collaborative spirit in open discussions, joint analyses, cross-consultations, case examinations, and reports with Jonas Petek, Claudio Nakayama, and Carlos Limberg. Similar appreciation goes to Daniel Zacher, Ricardo Takahira, and Alda Andrade from SAE Brazil for the opportunities to develop courses where automotive fire knowledge was put at the service of the engineering community.

Special appreciation goes to Ms. Ema Sutcliffe at EV FireSafe, Mr. Robert Braswell at the American Trucking Associations/Technology & Maintenance Council (TMC), Mr. Jeffrey Hyatt, Mr. Greg Sturdy at Carlisle Brake & Friction, Mr. Greg Lawrence at General Dynamics Land Systems, Mr. Victor Tsimpinos at Penske Truck Leasing Corporation, Mr. Frank Daniel Brubakken at Bridgehill, Mr. Russ Smolinsky at Amphenol PCD Shenzhen, B.F.A. Benjamin Grna-Hofstätter at Rosenbauer International AG, Shannon Solheim at Turtle Fire Systems, and all the companies, authors, and institutions that graciously granted permission to include crucial images and vital information in this book, ensuring its value for readers.

A heartfelt recognition goes to the selfless dedication of firefighters who bravely put their lives on the line daily, facing challenges created by vehicles, construction, unforeseen events, and acts of nature in their untiring commitment to saving lives and minimizing damage.

A special thank you is extended to Sherry Nigam, Amanda Zeidan, and the editorial team at SAE International, who believed in this book project, provided all the necessary means, and allowed it to become a reality.

Last but certainly not least, I want to express my wholehearted gratitude to my beloved daughters, Gi and Dani, whose invaluable reinforcement, proposed adjustments, and thoughtful feedback have significantly enhanced the coherence and scope of this book.

Introduction

Knowledge always desires increase; it is like fire, which must first be kindled by some external agent, but which will afterwards propagate itself.

—Samuel Johnson

In the world of transportation, automobiles have not only transformed how we navigate our world but also become an intrinsic part of our daily lives. From the early days of steam and electric vehicles to the dominance of internal combustion engine vehicles (ICEVs) for nearly a century, through the emergence of hybrid vehicles (HEVs) and the rebirth of battery electric vehicles (BEVs), automobiles have played a pivotal role in shaping our modern existence. However, with progress comes the challenge of addressing inherent risks, and one of the most critical concerns revolves around the threat of automotive fires.

This book explores the multifaceted landscape of automotive fires, unraveling the distinctive characteristics and safety challenges posed by diverse systems, architectures, and components of these vehicles. It is a comprehensive guide that delves into the underlying theories, preventive measures, and analytical techniques essential for understanding, mitigating, and investigating automotive fires.

As automotive technology evolves rapidly, the coexistence of traditional vehicles and their electrified counterparts introduces unique fire risks. In a deliberate effort to disambiguate terms, this book avoids the expression "electric vehicle" due to potential misunderstanding. Instead, the acronym BEV is used to refer to "pure" battery electric vehicles (i.e., vehicles without a combustion engine), and the abbreviation xEV is employed to encompass vehicles with an electrified powertrain (including hybrid electric vehicles, "pure" electric vehicles, and fuel cell vehicles).

Potential ignition sources and fuel loads vary from fuel system failures to battery thermal runaway events, necessitating an expanded knowledge base for effective fire safety measures. This book seeks to bridge this knowledge gap, addressing the specific challenges associated with each vehicle type.

The first section of this book (Theory) establishes the groundwork by elucidating the fundamental concepts underpinning automotive fires. Exploring the critical elements required for fire initiation, sustenance, and propagation delves into the associated thermal, chemical, mechanical, and electric mechanisms. This foundational understanding allows for the development of effective preventive measures and the proper analysis of real-life incidents. Concurrently, it examines ICEVs and xEVs, acknowledging their unique systems, components, and ongoing evolution, addressing their failure mechanisms, safety risks, challenges in fire combat, and evolving regulations.

The second section (Prevention) focuses on prevention strategies, drawing on industry best practices, safety standards, regulations, and advanced technologies. By exploring approaches to mitigate fire risks in both ICEVs and xEVs throughout the design, development, manufacturing, and maintenance phases, readers gain insights to safeguard their designs and vehicles effectively.

The third section (Analysis) centers on analyzing and investigating automotive fires. Utilizing real-world cases and expert insights, it delves into the methodologies used to determine the cause and origin of fires. Emphasizing thorough investigation techniques using the scientific method, this section underscores the crucial information gleaned from physical evidence, witness statements, and digital systems. Moreover, by examining the aftermath of automotive fires, invaluable lessons can be uncovered and inform future prevention strategies.

This book taps into the expertise of industry professionals, researchers, and fire safety specialists who have dedicated their careers to understanding, preventing, and analyzing automotive fires. Their invaluable insights, extensive research, and practical advice offer readers a holistic and authoritative perspective on this critical subject matter. It also examines real-world examples, providing practical instances to illustrate the concepts discussed.

I hope this book will serve as a comprehensive resource for professionals, researchers, and enthusiasts interested in understanding, preventing, and analyzing automotive fires. By bridging the gap between theory and practice, it aims to empower readers with the tools and insights necessary to safeguard lives, protect valuable assets, and pave the way for a secure and resilient automotive industry.

So, whether you are an automotive engineer, a fire safety professional, a forensic consultant, a fleet manager, an insurance investigator, or an enthusiast seeking to enhance your understanding of vehicle fires, this book serves as your indispensable guide.

Section 1: Theory

Chapter 01

Relevant Data/Statistics

1.1.
Introduction

This chapter explores the profound implications of car fires on a global scale. Along this book the term BEVs is used to denote electrified vehicles (EVs) without internal combustion engines (ICEs) and xEVs to denote a broader category encompassing hybrid vehicles (HEVs), battery-electric vehicles (BEVs), and fuel cell vehicles (FCEVs).

It surveys the intricate web of interconnected issues, from safety concerns and disruptions in road transportation to their influence on airport operations, maritime transport risks, insurance markets, and environmental consequences. The chapter also addresses manufacturing and design implications, technological advancements, and evolving safety standards.

Additionally, it examines global vehicle fire data, including statistics from the United States (US), the United Kingdom (UK), and other international perspectives. It explores the impact of car fires on ships and large parking structures, shedding light on transportation vessel challenges and incidents in expansive parking facilities.

As it unravels the area of origin, causes, and items first ignited, it considers the overall fire rates across vehicle types. The chapter further delves into recalls related to fire risk and the specific situations in which xEVs catch fire.

Embark on this concise exploration and navigation through the complex landscape of global car fires, uncovering their far-reaching consequences on safety, transportation, the environment, and the automotive industry.

1.2.
Global Relevance of Car Fires

Vehicle fires have worldwide importance for several reasons, although the impact may vary depending on the specific context and circumstances [1.1-1.6]. The following considerations underscore their widespread implications.

1.2.1.
Safety Concerns

Fire incidents involving vehicles can pose significant safety risks for the vehicle's occupants and nearby individuals and properties. Such an event in a crowded urban area or on a busy highway can lead to injuries, fatalities, and property damage.

1.2.2.
Road Transportation Disruptions

Fires involving vehicles can disrupt road transportation systems and infrastructure, particularly in urban areas, freeways, and major transportation hubs. Such disruptions lead to traffic congestion, delays, and logistical challenges, which have economic implications and affect the movement of goods and people.

1.2.3.
Impact on Airport Operations

Incidents involving fires in vehicle parking lots can have global consequences due to airports' critical role in international travel and commerce. Some of these events lead to disruptions in airport operations, impacting flights, passenger schedules, and cargo shipments [1.7-1.9].

1.2.4.
Maritime Transport Risks

Recent years have witnessed incidents involving fires on vehicle's transport ships, presenting risks to passengers, crews, and the transported vehicles. Some events have resulted in the total loss of vessels, with broader implications for the global automotive and transport businesses. These transport vessels, vital for intercontinental vehicle movement, can suffer significant financial losses, create supply chain disruptions, and cause environmental damage in the event of a fire. The impact extends beyond immediate safety concerns, affecting both the shipping companies and automotive manufacturers tied to the transported vehicles [1.10-1.15].

1.2.5.
Insurance Market Influence

Vehicle fires usually prompt insurance claims, and their increasing frequency or involvement of high-value vehicles can exert a global impact on insurance markets. Insurers may reevaluate policies and premiums, affecting individuals and businesses worldwide. Moreover, events on transport vessels or in large parking lots can result in substantial claims, straining insurance resources. The frequency and severity of these incidents influence how insurance companies approach coverage for vehicles in transit or parked in high-risk areas, shaping business practices on a broader scale.

1.2.6.
Environmental Consequences

Instances of car fires release pollutants and toxins, contributing to air, soil, and water pollution. Materials like plastics and chemicals in vehicles can generate hazardous emissions, influencing environmental concerns and regulations. Perceived fire risks in xEVs (whether realistic or not) can affect purchasing decisions and impact environmental plans to reduce emissions and greenhouse gases.

1.2.7.
Manufacturing and Design Implications

Incidents involving fires attributed to design flaws, manufacturing faults, or widespread issues with specific models can have global implications for the automotive industry. Manufacturers may face legal challenges, recalls, reputational damage, and financial strain, impacting their operations and market presence worldwide.

1.2.8.
Technological Advancements and Safety Standards

Fire occurrences can prompt advancements in vehicle safety technologies and the establishment of new or stricter safety standards [1.16, 1.17]. Global automotive trends often influence the adoption of safety features and regulations in various countries.

These examples demonstrate how car fires in specific contexts extend beyond local concerns, affecting international travel, trade, insurance, and global society.

1.3.
Global Vehicle Fire Data

The categorization and accessibility of vehicle fire data significantly differ across countries, presenting challenges for in-depth analysis. Some nations aggregate data for various vehicles, encompassing motorcycles, bicycles, scooters, buses, etc., while others lack official information. Focusing on a subset of available statistics, this section underscores the global concern of car fires [1.18-1.21]. It acknowledges the progress made by the automotive industry and authorities while emphasizing the necessity for additional global solutions.

1.3.1.
US

According to the National Fire Protection Association, approximately 188,500 fires of highway vehicles (i.e., motor vehicles intended for use on roadways) occurred in 2022, causing 610 civilian deaths, 710 civilian injuries, and $1.99

billion in direct property damage. **Figure 1.1** shows the yearly evolution of the fires for highway vehicles based on data from the same source. While there was a significant reduction from 1990 to 2010, the trend has stagnated around 177,000 fires per year. Identifying the precise causes of these divergent trends exceeds the scope of this book.

Figure 1.1 Highway vehicle fires in the US.

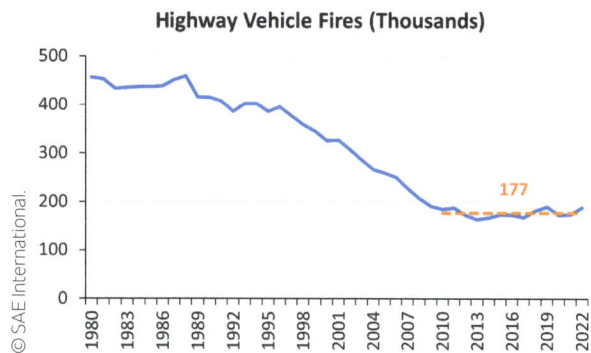

Various contributing factors have enhanced the situation. These include the reduction in the number of accidents per capita; a decrease in overall vehicle failure rates; advancements in regulations, crash tests, and recalls; and the continuous technological evolution of vehicles, encompassing improvements in safety systems, parts, and materials. The recent stagnation could be linked to the growing complexity of vehicle electronics and systems, coupled with the learning curve associated with the widespread adoption of new technologies.

1.3.2.
UK

According to the British Home Office fire statistics, England experienced 19,135 road vehicle fires in 2022/2023, causing 22 fatalities and 460 nonfatal casualties.

1.3.3.
Other International Data

The International Association of Fire and Rescue Services (CTIF) publishes the World Fire Statistics

Magazine, which summarizes fire incident data for various countries and cities. Unfortunately, *car* fire data are unavailable for most countries, resulting in a limited vision of the global situation. The countries reporting the highest vehicle fire numbers in 2021 are listed here:

- US: 208,500
- France: 45,170
- Russia: 17,249
- Poland: 9654
- Japan: 3512

It is crucial to clarify that the US data presented in this section pertain to vehicles in general, extending beyond highway vehicles covered in the preceding section of this chapter. Additionally, the data are sourced from different outlets. For a detailed examination of the diverse data sources within the US, readers are encouraged to explore the technical study titled "Vehicle Fire Data: Different Sources, Different Goals, Different Conclusions?" by R. Ray and M. Ahrens [1.20].

1.4.
Car Fires in Ships and Large Parking Structures

While most car fires occur on open roads, incidents in large vehicle transport ships and expansive parking structures result in substantial economic losses and heightened media attention, prompting evolving regulations. Amid the transition from ICEVs to xEVs, there is a swift inclination to attribute these incidents to "electric vehicles." However, some of the incidents were triggered by ICEVs, and preliminary data reveals that, in fact, the number of BEV fires is much smaller than ICEV fires.

It must be pondered that ICEVs make more intense use of plastics now than they did a few

decades ago. This increases the magnitude and duration of the fire event as more combustible materials are available. Moreover, plastic fuel tanks became the norm. Although regulations demand them to resist external flames for a few minutes, beyond this time, they are likely to leak and pour their content, feeding and enlarging the fire. Ancient metallic tanks generally avoided (or at least delayed) this type of event.

The heightened plastic content in modern vehicles not only accelerates the propagation of flames but also expedites ignition and the expansion of fires to nearby vehicles within densely packed transportation vessels and parking structures. This poses a significant challenge for fire control, especially when local measures such as sprinkler systems prove inadequate in containing the initial vehicle fire.

In such scenarios, firefighting crews arriving at the scene face a complex situation where the fire has already extended to multiple vehicles. This complicates the process of extinguishing the fire, particularly in expansive parking structures, where the efforts may extend beyond a day. Moreover, in the case of large vehicle transport vessels, the extinguishing process might span more than a week due to the unique challenges posed by these environments. These challenges necessitate a strategic and comprehensive approach to firefighting efforts to effectively tackle and mitigate the escalating threats in these high-risk settings.

1.4.1.
Transportation Vessels

In addition to the challenges posed by tight packing and restrained access, ample car transportation ships face a unique issue during fires—excessive water application may cause stability problems, potentially tipping the boat.

The following sections provide examples of notable incidents involving large ships.

1.4.1.1.
Sincerity Ace (December 2018)
On New Year's Eve 2018, the Sincerity Ace caught fire carrying 3500 cars in the Pacific Ocean. The crew abandoned the ship, with 16 rescued and five presumed dead. The cause remains unknown, with some sources suggesting a possible link to the transportation of "electric vehicles." The fire burned for about ten days, and the vessel was eventually towed to Japan for salvage.

1.4.1.2.
Felicity Ace (February 2022)
The Felicity Ace was transporting nearly 4000 cars, including luxury brands, when it encountered an eight-day fire on February 16, 2022. The crew abandoned the ship, and all 22 members were rescued by a nearby tanker. While the cause is unknown, some sources speculate a potential connection to the transportation of "electric vehicles." The ship sank on March 1, 2022, in the North Atlantic.

1.4.1.3.
Fremantle Highway (July 2023)
The Fremantle Highway, carrying 3783 new cars, including 498 "electric vehicles," caught fire off the Dutch coast on July 25, 2023. One sailor died, and 22 were rescued. The intense fire damaged the upper decks of the ship, as seen in **Figure 1.2**, causing some cars to melt together. Approximately 2700 vehicles were destroyed, and the boat was towed for salvage. An "electric vehicle" is suspected to have triggered the deadly blaze, which took about ten days to extinguish.

Figure 1.2 Freemantle Highway fire.

Michelle Holton/Shutterstock.com.

1.4.2.
Car Fires in Large Parking Structures

Similar to incidents involving large vessels, fires in expansive parking structures are often initially attributed to "electric vehicle" failures. With many parking facilities equipped with surveillance cameras, subsequent investigations frequently reveal that an ICEV was the actual cause.

The following sections provide examples of incidents in these structures [1.22-1.24].

1.4.2.1.
Liverpool Echo Arena (December 2017)

A fire ignited when an ICEV sports utility vehicle (SUV) burst into flames, spreading to other vehicles and destroying 1400 cars. While there were no injuries, the car park suffered severe damage, as seen in **Figure 1.3**, and required demolition.

Figure 1.3 Liverpool Echo Arena parking fire.

Ken Biggs/Alamy Stock Photo.

1.4.2.2.
Norway Stavanger (Sola) Airport (January 2020)

A significant parking garage fire at an airport on the west coast of Norway destroyed approximately 300 cars, disrupting air traffic and leading to facility evacuations. The parking complex structures partially collapsed. Initially reported as an "electric vehicle" fire, later investigations revealed that a 2005 diesel ICEV with wiring problems sparked the blaze.

1.4.2.3.
London Luton Airport (October 2023)

The fire at Luton Airport car park on October 10, 2023, destroyed over 1000 cars and caused significant structural collapse on the third floor. Initially attributed to an "electric vehicle" fire, investigators later determined that a diesel ICEV SUV was responsible for starting the fire.

1.5.
Area of Origin, Cause, and Item First Ignited

Figures 1.4, 1.5, and 1.6 offer a comprehensive insight into highway vehicle fires, utilizing data from 2013 to 2017 from the National Fire Protection Agency. Figure 1.4 details the origin area of vehicle fires, emphasizing the engine, running gear, or wheel as the primary areas. Figure 1.5 delves into the key ignition causes, with unintentional incidents and failures or heat sources emerging as the predominant factors. Figure 1.6 highlights the primary items initially ignited, with electrical components and combustible fluids being more frequently involved. Each category is presented as a percentage of occurrence. It is noteworthy that considering the data collection time frame, these incidents predominantly pertain to events associated with ICEVs.

Figure 1.4 Percentage of fires by area of origin.

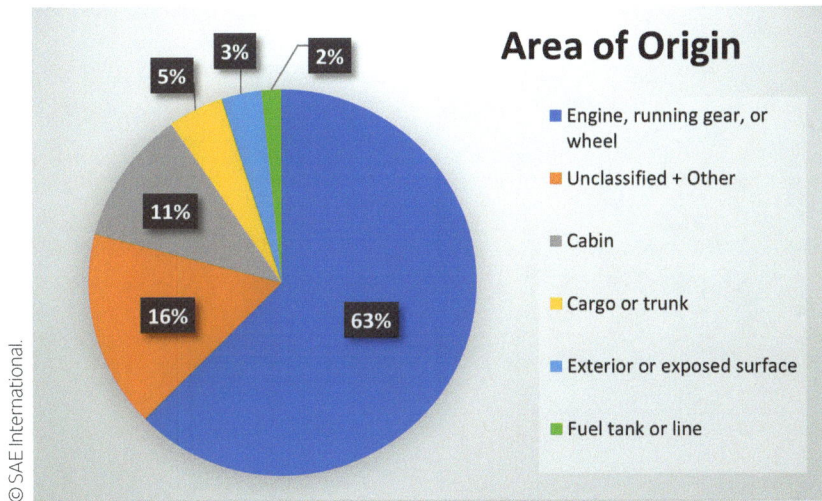

© SAE International.

Figure 1.5 Percentage of fires by cause of ignition.

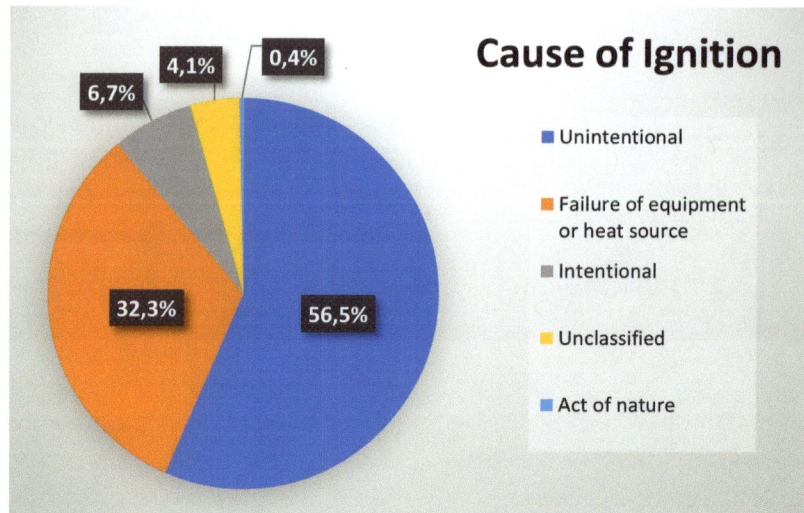

© SAE International.

Figure 1.6 Percentage of fires by item first ignited.

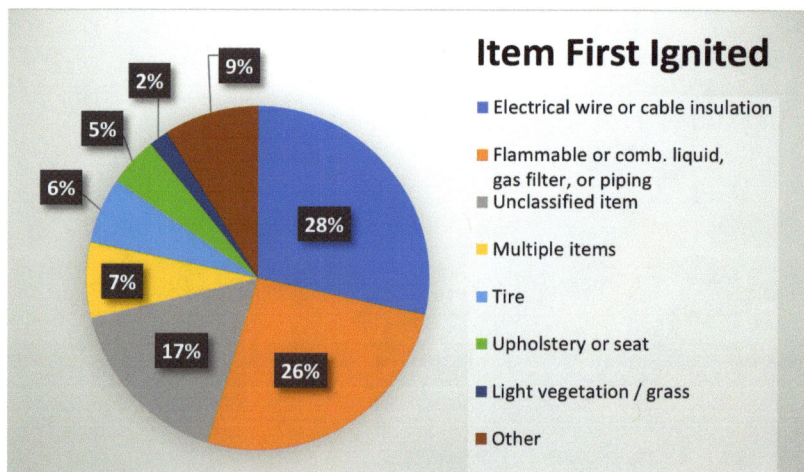

© SAE International.

1.6.
Car Fire Rates by Type

When a traction lithium battery ignites, the spectacle is remarkable and poses a more significant challenge for authorities to contain. This has resulted in widespread misinformation regarding the frequency and severity of fires in xEVs.

Despite the heightened media attention, the total number of annual fires in xEVs in the US remains considerably lower than those in ICEVs. Data compiled by AutoinsuranceEZ from the National Transportation Safety Board (NTSB) and Bureau of Transportation Statistics (BTS) for 2021, used as reference for **Table 1.1**, revealed 199,533 fires in ICEVs, 16,051 in HEVs, and 52 in BEVs. When these numbers are translated into a proportion of vehicles sold that year, the Pareto analysis demonstrates that HEV fires have the highest frequency to sales (approximately 3475 per 100k sales), followed by ICEVs at approximately half the frequency (1530 per 100k sales). In contrast, BEVs have a much lower incidence (25 per 100k sales) [1.25-1.28].

Table 1.1 Fires and fire rates by vehicle type.

Vehicle type	Fires (occurrences)	Fire rate (fires/100k sales)
ICEV	199,533	1530
HEV	16,051	3475
BEV	52	25

© SAE International.

1.7.
Recalls Related to Fire Risk

AutoinsuranceEZ also compiled 2020 data on recalls related to fire risks [1.25]; this information is summarized in **Table 1.2**. The volume of ICEVs susceptible to catching fire, if not addressed (1,085,800), was notably higher than the numbers for BEVs (152,000) and HEVs (32,100). While there is not a precise correlation between the quantity of vehicles recalled and the actual number of fires, the Pareto analysis emphasizes the more significant number of ICEVs that were carrying a known fire risk.

Table 1.2 US recalls related to fire risk.

Vehicle type	Quantity recalled
ICEV	1,085,800
BEV	152,000
HEV	32,100

© SAE International.

1.8.
xEV Situation when Fire Started

An important statistic is the proportion of xEV fires to the vehicle's situation when the fire began. **Figure 1.7** presents a pie chart derived from a recent study on lithium-ion battery (LIB) fires conducted by Bisschop et al., encompassing incidents from 2010 to 2019. The notable proportion of fires attributed to immersion is influenced by 19 incidents linked to the flooding caused by Hurricane Sandy in 2012. Excluding those, no predominant vehicle status appears more likely to initiate a fire. This underscores the need for improvements in various directions to reduce the occurrence of xEV fires [1.29, 1.30].

Figure 1.7 xEV situation when the fire started.

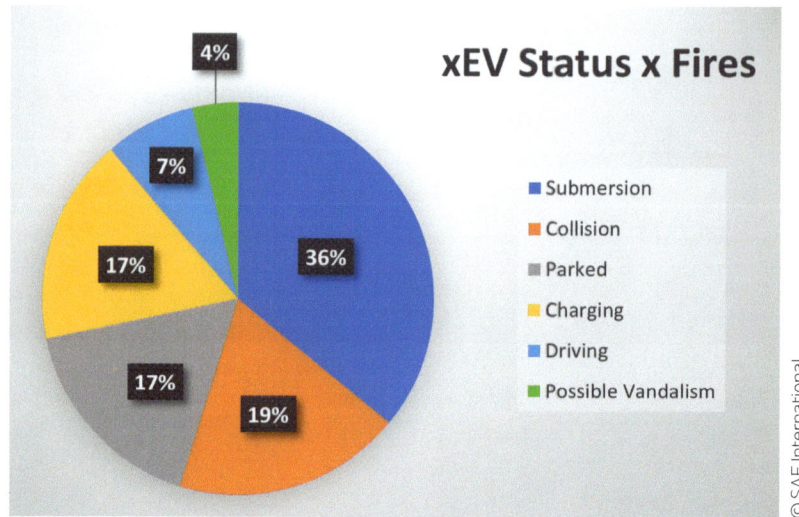

© SAE International.

References

[1.1]. Hall, S., "Fire Loss in the United States," National Fire Protection Association Research, November 2023, accessed January 2024, https://www.nfpa.org/education-and-research/research/nfpa-research/fire-statistical-reports/fire-loss-in-the-united-states.

[1.2]. Home Office from the UK Government, "FIRE0302: Primary Fires, Fatalities and Non-Fatal Casualties in Road Vehicles by Motive and Vehicle Type, England," accessed January 2024, https://assets.publishing.service.gov.uk/media/64be64401e10bf000e17ccef/fire-statistics-data-tables-fire0302-270723.xlsx.

[1.3]. International Association of Fire and Rescue Services (CTIF), "World Fire Statistics," Number 28, 2023 (2021 data), accessed January 2024, https://ctif.org/sites/default/files/2023-06/CTIF_Report28-ESG.pdf.

[1.4]. The National Fire Data Center (NFDC), "Fire in the United States—2008-2017," 20th Edition, US Fire Administration (USFA), Federal Emergency Management Agency (FEMA), 2019.

[1.5]. The National Fire Data Center (NFDC), "Highway Vehicle Fires (2014-2016)," Topical Fire Report Series, Volume 19, US Fire Administration (USFA), Federal Emergency Management Agency (FEMA), 2018.

[1.6]. US Fire Administration (USFA), Federal Emergency Management Agency (FEMA), "Fire in the United States 2008-2017," 20th Edition, November 2019.

[1.7]. Automotive News Europe, "Hundreds of Cars Destroyed by Fire at Norway Airport," January 8, 2020, accessed January 2024, https://europe.autonews.com/automakers/hundreds-cars-destroyed-fire-norway-airport.

[1.8]. Itv News, "Luton Airport Fire: Cars on First Three Floors of Carpark Will Not Be Recovered," November 3, 2023, accessed January 2024, https://www.itv.com/news/anglia/2023-11-03/cars-on-first-three-floors-of-burnt-out-luton-airport-carpark-not-salvageable.

[1.9]. Storesund, K., Sesseng, C., Bøe, A., and Mikalsen, R., "Investigation of Massive Fire in a Multi-Storey Car Park in Norway," RISE Fire Research, Norway, accessed January 2024, https://www.ri.se/sites/default/files/2020-12/FRIC%20D1.2-2020_01%20FIVE%20conference%20presentation%20Multi-storey%20car%20park%20fire%2C%20presentation.pdf.

[1.10]. Boehmer, H., Klassen, M., and Olenick, S., "Modern Vehicle Hazards in Parking Garages and Vehicle

Carriers," Fire Protection Research Foundation, National Fire Protection Association, July 2020, accessed January 2024, https://www.nfpa.org/education-and-research/research/fire-protection-research-foundation/projects-and-reports/modern-vehicle-hazards-in-parking-garages-vehicle-carriers.

[1.11].   CBC News, "2 Stranded Truck Drivers Pulled from Burning Ferry Near Greek Island, 11 Still Missing," February 18, 2022, accessed January 2024, https://www.cbc.ca/news/world/greece-ferry-fire-1.6356393#:~:text=The%20cause%20of%20the%20blaze%20was%20unclear.%20The,off%20the%20island%20of%20Corfu%2C%20Greece%2C%20on%20Friday.

[1.12].   Det Norske Veritas (DNV-GL), "Fires on Ro-Ro Decks," Paper No. 2016-P012, April 2016, accessed February 2024, https://www.dnv.com/news/enhancing-fire-safety-on-ro-ro-decks-69059.

[1.13].   EV Fire Safe, "Electric Vehicle Fires on Ships & Ferries," EV Fire Safe Is Supported by the Australian Government, Department of Defence, accessed January 2024, https://www.evfiresafe.com/post/electric-vehicle-fires-on-ships-ferries.

[1.14].   Financial Post, "500 Electric Vehicles on Board Burning Ship Off Netherlands," July 28, 2023, accessed January 2024, https://financialpost.com/commodities/energy/electric-vehicles/500-evs-burning-ship-netherlands.

[1.15].   Greek City Times, "There Were Drivers Sleeping in Their Trucks: Shocking Photos and Testimonies from Euroferry Olympia Blaze," February 18, 2022, accessed January 2024, https://greekcitytimes.com/2022/02/18/euroferry-olympia-corfu/.

[1.16].   O'Connor, B., "Fire Safety for Electric Vehicles and Other Modern Vehicles in Parking Structures," National Fire Protection Association, November 2022, accessed January 2024, https://www.nfpa.org/news-blogs-and-articles/blogs/2022/11/28/evs-and-parking-structures.

[1.17].   Underwriters Laboratories (UL), "Electric Vehicle Fire Data and Concerns for First and Second Responders," Battery Safety Science Webinar, Presented by Robert "Bob" Swaim, April 19, 2021, accessed January 2024, https://ul.org/sites/default/files/2021-04/UL_FF_Issues_20210416WEB.pdf.

[1.18].   Ahrens, M., "Vehicle Fires," National Fire Protection Association, March 2020, accessed January 2024, https://content.nfpa.org/-/media/Project/Storefront/Catalog/Files/Research/

NFPA-Research/US-Fire-Problem/osvehiclefires.pdf?rev=ce9308b1447140ef9bef693635a96d71.

[1.19].   Ahrens, M., "Vehicle Fires—Supporting Tables," National Fire Protection Association, March 2020, accessed January 2024, https://content.nfpa.org/-/media/Project/Storefront/Catalog/Files/Research/NFPA-Research/US-Fire-Problem/osvehicle-firestables.pdf?rev=28fd79a2aed347f0b4f1766ef53bbd34&hash=52CF58909C18DDBA81E8E63BE5D52414.

[1.20].   Ray, R. and Ahrens, M., "Vehicle Fire Data: Different Sources, Different Goals, Different Conclusions?" SAE Technical Paper 2007-01-0877 (2007), doi:https://doi.org/10.4271/2007-01-0877.

[1.21].   US Fire Administration (USFA), Federal Emergency Management Agency (FEMA), "Causes of Vehicle Fires (2021)," accessed January 2024, https://www.usfa.fema.gov/statistics/outside-vehicle-fires/.

[1.22].   Australia News, "Multiple Vehicles Destroyed after Electric Car's Lithium-Ion Battery Sparks Carpark Fire at Sydney Airport," September 12, 2023, accessed January 2024, https://www.skynews.com.au/australia-news/multiple-vehicles-destroyed-after-electric-cars-lithiumion-battery-sparks-carpark-fire-at-sydney-airport/news-story/232eb3f7fe6e8b723309d1edd3ed52d2.

[1.23].   BBC News, "Ferocious Fire Ripped through Liverpool Echo Arena Car Park," January 1, 2018, accessed January 2024, https://www.bbc.com/news/uk-england-merseyside-42533830.

[1.24].   Grossman, H., Ray, R., Zhao, K., and Kytömaa, H., "Analysis of Garage Fires," SAE Technical Paper 2006-01-0791 (2006), doi:https://doi.org/10.4271/2006-01-0791.

[1.25].   AutoinsuranceEZ, "Gas vs. Electric Car Fires [2024 Findings]," December 2023, accessed January 2024, https://www.autoinsuranceez.com/gas-vs-electric-car-fires/.

[1.26].   EV Fire Safe, "Passenger EV LIB Fire Incidents," EV Fire Safe Is Supported by the Australian Government, Department of Defence, accessed January 2024, https://www.evfiresafe.com/_files/ugd/8b9ad1_01aa449ee5074086a55cb42aa7603f40.pdf.

[1.27].   Inside EVs, "Not an Electric Car Fire: Diesel Opel Responsible for Airport Blaze," January 9, 2020, accessed January 2024, https://insideevs.com/news/392047/bloomberg-ev-fire-cause-diesel/.

[1.28].   The Sunday Times – Driving, "Are Electric Cars More Likely to Catch Fire Than Petrol and Diesel Cars? Commentators Are Quick to Blame EVs," November 20, 2023, accessed January 2024, https://www.driving.co.uk/car-clinic/are-electric-cars-more-likely-to-catch-fire-than-petrol-and-diesel-cars/#:~:text=In%20fact%2C%20independent%20research%20into,fire%2C%20there%20are%20unique%20challenges.

[1.29].   Sun, P., Bisschop, R., Niu, H., and Huang, X., "A Review of Battery Fires in Electric Vehicles," *Fire Technology* 56 (2020): 1361-1410, doi:https://doi.org/10.1007/s10694-019-00944-3.

[1.30].   Chang, C.-H., Gorin, C., Zhu, B., Beaucarne, G. et al., "Lithium-Ion Battery Thermal Event and Protection: A Review," *SAE Int. J. Elec. Veh.* 13, no. 3 (2024): 1-41, doi:https://doi.org/10.4271/14-13-03-0019.

Chapter 02

Automotive Fire Science Fundamentals

2.1. Introduction

In this book, the term "fire" primarily refers to uncontrolled combustion within a vehicle or its components, which can cause significant damage. This type of combustion is a chemical reaction that releases energy in the form of heat and light, often producing by-products such as water vapor, oxides, and smoke. It is important to note that while this definition is used throughout most of the book, the term "fire" can have different meanings in other contexts, including within the automotive industry itself.

It is crucial to have a basic understanding of fire theory to comprehend the behavior of automotive fires. Although this chapter covers only some technical aspects of fire science, the concepts offered in it provide a foundation for understanding the various factors contributing to the formation and spread of fire and the techniques utilized to prevent and extinguish it [2.1-2.3].

2.2. Fire Triangle

The fire triangle, also known as the combustion triangle and sketched in **Figure 2.1**, is a simple bidimensional model that depicts the three elements required to initiate a fire: fuel, energy source, and oxidizer. Usually, the diagram shows oxygen as the oxidizer and heat as the energy source. Either way, a fire cannot start without any of these three elements.

Figure 2.1 Fire triangle.

Ozant/Shutterstock.com.

Figure 2.2 Fire tetrahedron.

BALRedaan/Shutterstock.com.

In this model, fuel is any material that can be burned, such as paper or gasoline, and it is not restricted to the fluids usually employed to propel a vehicle. An energy source, such as a spark or a flame, is often required to initiate combustion. Oxygen is the most frequent oxidizer, modifying the fuel substances at a molecular level through chemical reactions that release energy in the form of heat and light.

The fire triangle is illustrated as a three-sided figure, each representing one of the elements required for a fire to start. The triangular shape emphasizes the interconnectedness of these elements and the fact that they must be present in correct proportions to initiate a fire.

2.3.
Fire Tetrahedron

While the fire triangle is a helpful model for understanding the basic requirements for a fire to *start*, it does not capture the complex chemical reactions required to *sustain* the combustion. The fire tetrahedron model, depicted in **Figure 2.2**, expands on the fire triangle by adding a fourth element: chemical chain reaction.

2.3.1.
Chain Reaction

The chain reaction is a process by which the heat produced along the oxidation of the fuel elevates the temperature of surrounding amounts of oxidizer and fuel, allowing more molecules to react. Moreover, the chain reaction usually encompasses the generation of other (intermediate) molecules, some capable of exothermic reactions with the oxidizer. Exothermic reactions release energy in the form of heat to the surroundings.

Once combustion has initiated, the heat released by the oxidation reaction might allow more molecules to oxidize until the oxidizer or fuel is consumed. In fact, at the molecular level, several intermediate reactions occur, with some being exothermic and providing the necessary heat for the oxidation of other substances and the creation of by-products to continue.

The chain reaction is a critical component of fire because it ensures that the energy released during combustion is continually available to sustain the fire. Therefore, it is depicted at the base of this pyramidal model.

2.3.2.
Free Radicals

At the molecular and atomic levels, the chain reaction becomes significantly more intricate as it involves the generation of free radicals. Free radicals are highly reactive chemical entities with unpaired electrons, rendering them inherently unstable and short-lived.

When a combustion reaction initiates, the application of activation energy breaks certain atomic bonds within fuel molecules, generating free radicals such as hydroxyl and alkyl radicals. Some of these will react intensely with oxygen in the air, releasing significant heat. At the same time, others will continue to break down additional fuel molecules, thereby generating more free radicals. Finally, the remaining free radicals will combine with other atoms or molecules, forming more stable chemical species.

The chain reaction is a simplified representation of this complex process. It highlights that given fuel, oxidizer, and heat, we obtain a self-sustaining activity that produces more heat and chemical by-products.

2.4.
Combustible Materials/Fuels

Combustible materials ("fuels" in the usual fire triangle and tetrahedron models) generally have a high carbon or hydrogen content, such as gasoline and propane. Many can easily ignite and sustain a flame. Conversely, materials with a low carbon or hydrogen content, such as metals and ceramics, are generally not combustible at ambient temperature, and several of these are fire-resistant.

In current automobiles, combustible materials are present in several states and forms everywhere, from bumper to bumper: oils in the engine compartment, fluids in the fuel systems, miscellaneous plastic parts, to name a few.

Combustible materials can appear in solid form, such as plastics, upholstery, foams, labels, dried paint, and even metals. Once ignited, magnesium alloys (often used in the steering wheel) and lithium (in the metallic form in some traction batteries) burn vigorously.

Some examples of combustible materials in liquid form are gasoline, ethanol, diesel, oils, and hydraulic fluids. Various are in the transmission, brake, lubrication, and other hydraulic systems. Moreover, certain automotive liquids do not readily burn at ambient temperature, requiring at least moderate heating to be ignitable.

Some combustibles might be present in the gaseous form, either directly stored as such (e.g., hydrogen), after depressurization (such as liquefied petroleum gas [LPG]), or as vapors when particular liquids are heated (as it happens with diesel and several oils).

Combustibles are classified as volatiles when their vapors are flammable at ambient temperature and pressure. Gasoline and ethanol belong to that category, while diesel and most automotive oils are not volatile [2.4-2.8].

2.5.
Oxidizer

This is the substance that reacts with the fuel (or its gases) and forms a flame. In chemical terms, this substance gains electrons along the oxidizing (or reduction) reaction. Its quantity regulates the flame intensity, and high amounts of oxidizing agents can generate explosions.

The most common oxidizer, oxygen, naturally occurs in the gaseous form. However, other oxidizers exist, several in different physical states. Potassium nitrate is a solid oxidizer that allows gunpowder to burn violently. Hydrogen peroxide, in the liquid state, dramatically oxidates potassium iodide in a notorious mid-school laboratory example.

2.6.
Energy Source

Usually, the simple combination of fuel and oxidizer does not start a fire. The role of the energy source is to provide enough heat (activation energy) at a molecular level for the oxidation reaction to begin. From this perspective, it is unnecessary to elevate the temperature of the complete fuel and oxidizer volumes; heating or augmenting the pressure of a few molecules in the right proportion might be sufficient for the fire to start. This can be implemented by several physical mechanisms: a flame, an ember, an impact, a shock wave, enough pressure, friction, infrared irradiation, a mechanical or electrical spark, arcing, or lightning. Some physical mechanisms are illustrated in **Figures 2.3** through **2.7**, while Chapters 3, 4, and 5 will examine energy and degradation sources in more detail. Relevant in the automotive environment is that electric

energy, through the Joule effect and arcing, can elevate the temperature to the point where combustion will start [2.9].

Figure 2.3 Lit match.

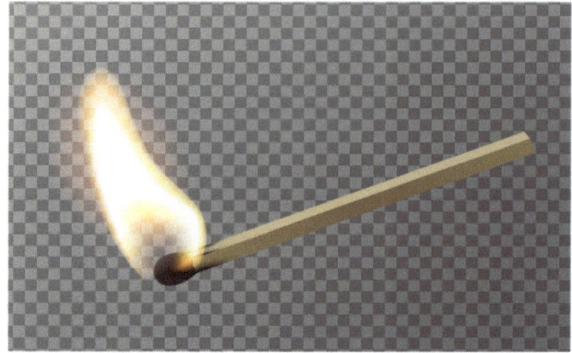

backUp/Shutterstock.com.

Figure 2.4 Embers in a smoldering campfire.

denisik11/Shutterstock.com.

Figure 2.5 Hammer impact fires the primer.

Frunze Anton Nikolaevich/Shutterstock.com.

Figure 2.6 Red glowing disc brake.

BLKstudio/Shutterstock.com.

Figure 2.7 Spark plug firing.

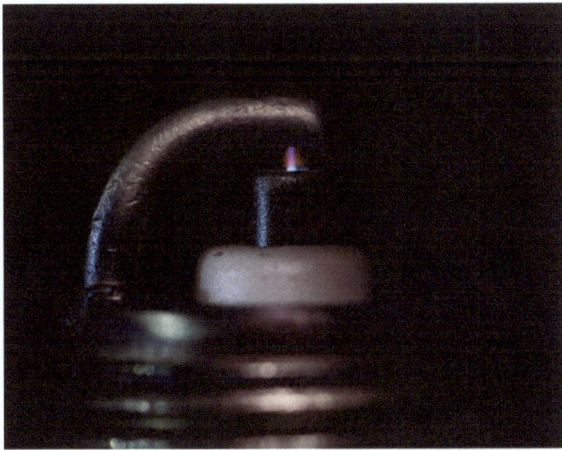

J.J. Gouin/Shutterstock.com.

2.7.
Explosion

An explosion occurs when the rate of energy released by combustion is faster than the rate at which the gases can escape, causing a rapid increase in pressure. This pressure buildup bursts enclosures and creates a shock wave that propagates through the surrounding air or materials, often causing damage to structures and objects in its path. On the other hand, it purposely happens in combustion engines, allowing us to harvest the mechanical energy that displaces the pistons.

2.8.
Airbags

The inflation of airbags is achieved through a controlled explosion that does not involve any combustion or use of fuel and oxidizer. Instead, a small amount of sodium azide (NaN_3) is utilized. When heated, this compound rapidly decomposes in an exothermic reaction, producing nitrogen gas and metallic sodium. It is the nitrogen gas that rapidly inflates the airbag. An electronic module generates an ignition spark that triggers the chain reaction. The operation of one of these systems is portrayed in **Figure 2.8**.

Figure 2.8 Airbag deflagration.

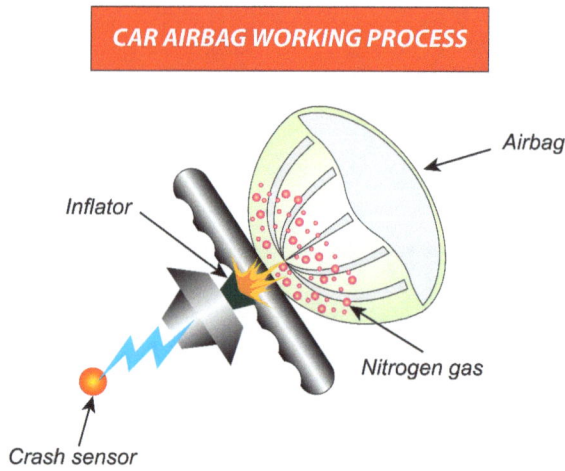

CAR AIRBAG WORKING PROCESS

Airbag

Inflator

Nitrogen gas

Crash sensor

Vadober/Shutterstock.com.

Some automotive examples at ambient pressure are given in **Table 2.1**. These temperatures are approximate laboratory values as the exact composition of these automotive fluids changes from country to country and might also change over time due to the evolution of local regulations. Moreover, in real life, some thresholds depend on the surrounding environment (e.g., the shape, composition, and surface roughness of a metallic part that heats the fuel vapors and the local ventilation conditions). The same holds for the fire point and autoignition temperatures listed in the same table.

2.9.
Flash Point

The flash point is the lowest temperature, at ambient pressure, at which a fuel gives off enough vapors to create an ignitable mixture with the air surrounding its surface. An external heat source ignites the combustion, but the reaction is *not sustained* once this heat source is removed, as illustrated in **Figure 2.9**.

Table 2.1 Typical flash point, fire point, and autoignition temperature for automotive fluids.

Substance	Flash point °F (°C)	Fire point °F (°C)	Autoignition temperature °F (°C)
Diesel	100–205 (38–96)	104–212 (40–100)	350–625 (176–329)
Ethanol	57 (13)	61 (16)	685 (363)
Gasoline	−45 (−43)	−40 (−40)	536–700 (280–371)
Engine oil	300–500 (149–260)	302–509 (150–265)	500–700 (260–371)

© SAE International.

2.10.
Fire Point

The fire point, also known as the combustion point, is the temperature at which a fuel gives off enough vapors to be ignited by an external heat source and *continue to burn* even when this heat source is removed. **Figure 2.10** exemplifies what would happen in our laboratory setup. The fire point is typically just a few degrees higher than the flash point.

Figure 2.9 Laboratory experiment—flash point.

Maji Design, Mis Dsgn, and VectorArtist7/Shutterstock.com.

Figure 2.10 Laboratory experiment—fire point.

Maji Design, Mis Dsgn, and VectorArtist7/Shutterstock.com.

2.11.
Autoignition Temperature

Autoignition temperature is the temperature required to ignite, *without an external energy source*, the vapors emerging from a fuel. **Figure 2.11** represents the events in the same laboratory setup. The autoignition point is usually significantly higher than the flash and fire points. In the automotive environment, it justifies, for instance, the ignition of gasoline droplets coming from a leaky fuel line if they reach a hot exhaust manifold. However, the real-world autoignition temperature can be influenced by various factors [2.10-2.13].

Figure 2.11 Laboratory experiment—autoignition temperature.

Maji Design, and Mis Dsgn/Shutterstock.com.

2.12.
Flammability Classifications

The term "flammability" refers to how easily a material will ignite and sustain combustion. Generally, flammable materials can be easily ignited and burn rapidly. Nonflammable materials, on the other hand, do not ignite easily and do not sustain a fire. More detailed classifications and test methods exist according to standards such as SAE J369, FMVSS 302, ISO 3795, and UL-94. These classifications consider factors such as the material's capability of being ignited, horizontal and vertical fire spread rates, self-extinguishing properties, and the release of flaming drops or particles [2.14-2.17].

Several factors can affect a material's flammability, including its chemical composition, moisture content, and surface area. For example, materials containing volatile compounds or dry and crumbled materials are more likely to ignite and burn than wet materials containing nonvolatile compounds.

Although it is not technically or economically feasible to impose that all materials used in an automobile are not flammable, it is possible to define acceptable flammability for a given application from an engineering perspective. Fortunately, many flame-retardant additives have been developed to improve the characteristics of polymers and insulation materials. These additives contribute to advancements in safety, performance, and cost-effectiveness.

2.13.
Flammability Range

The range of fuel vapor concentrations in the air that can sustain combustion is known as the flammability range of that fuel. Below the lower end of the range, there is not enough fuel vapor to support combustion, while above the upper end, the mixture is too rich to ignite. The range between these two limits is also called the explosive range.

The lower flammability limit (LFL) is the minimum vapor concentration in air below which the fuel–air mixture is too lean to ignite. The upper flammability limit (UFL) is the maximum concentration of vapor in air above which there is not enough oxygen to support combustion.

For gasoline vapors, the flammability range goes approximately from 1.4 to 8.0%. For hydrogen, as a comparison, the range is much more extensive: from about 4 to 75%.

2.14.
Heat Transfer: Conduction, Convection, and Radiation

The exchange of thermal energy between different bodies or systems due to a temperature difference is known as heat transfer. As shown in **Figure 2.12**, there are three main modes of heat transfer: conduction, convection, and radiation.

Figure 2.12 Heat transfer mechanisms.

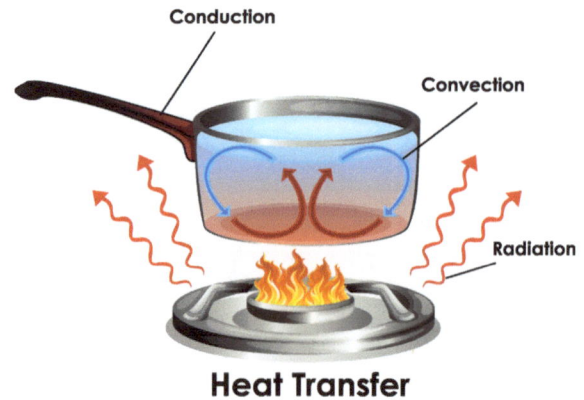

BlueRingMedia/Shutterstock.com.

2.14.1.
Conduction

Conduction is the transfer of heat between two objects in contact with each other or through a continuous material by transferring energy from one molecule to another. This is typically seen in solids, such as metals, where heat is conducted from one end of the object to another.

2.14.2.
Convection

The process of transferring heat energy through the movement of a fluid, such as air or water, is called convection. When a portion of the fluid is heated, its molecules gain power, causing them to vibrate more intensely and spread out. Consequently, the volume they occupy expands, leading to a decrease in density. As a result, the heated portion rises, while cooler, denser portions of the fluid sink and take its place. This molecule flow establishes a heat transfer cycle within the fluid and its surroundings, generally transferring the heat faster than what would be achieved through conduction alone.

2.14.3.
Radiation

Radiation is the process of heat energy transfer using electromagnetic waves. Unlike conduction and convection, which rely on a medium, radiation can occur even in a vacuum and is the fastest heat transfer mechanism. It happens when charged particles move within atoms and molecules, generating electromagnetic waves. In the context of fire theory, a relevant example is the infrared light emitted by hot surfaces.

2.15.
Pyrolysis and Carbonization

Pyrolysis and carbonization are two thermal processes that are closely related. They involve the decomposition of materials at high temperatures in the absence or near-absence of oxygen.

2.15.1.
Pyrolysis

Pyrolysis breaks substances down into smaller, simpler molecules. During pyrolysis, the material is heated to several hundred degrees Fahrenheit, causing the bonds between the atoms to break. This often releases shorter molecules in the form of volatile gases and liquids and leaves behind solid residues called "char." Frequently, some of these by-products are flammable, fueling combustion reactions. The burning of tires, exemplified in **Figure 2.13**, involves intricate pyrolysis reactions of the original rubber and its subproducts.

Figure 2.13 Tires burning.

Joa Souza/Shutterstock.com.

2.15.2.
Carbonization

Carbonization, in turn, is a particular type of pyrolysis that refers to converting organic materials into carbon-rich materials, such as charcoal or coke. Carbonization typically involves heating the organic material to a higher temperature than in standard pyrolysis, usually above 750°F (400°C), and for a more extended heating period. Volatile gases and liquids are driven off during carbonization, leaving a porous, carbon-rich material behind. Of particular importance in electronics, many plastic insulators, once carbonized, will conduct electric current. This might allow a significant current flow, which will cause a further rise in temperature due to the Joule effect. Carbonization of printed circuit boards, as exemplified in **Figure 2.14**, leads to significant damage and fire risks.

2.16.
Fluid Leakages

Automobiles contain many fluids that can undergo combustion. While the design and production of vehicles effectively confine these fluids within their reservoirs and distribution lines, their integrity may not be sustained over time. Incidents such as collisions can rupture fuel lines, the aging of components can compromise seals, and inadequate maintenance practices can lead to oil leaks, among other examples.

A fire can ignite if these flammable liquids or their vapors encounter an appropriate ignition source, such as a hot exhaust manifold or a high-voltage spark. Furthermore, in the event of an existing fire, the leakage of these fluids can introduce additional fuel, intensifying and prolonging the fire. In cases where the fluid is

Figure 2.14 Carbonized electronic board.

Doug McLean/Shutterstock.com.

pressurized, the leaked fluid can cause flames to spread significantly from the point of escape. Consequently, this can expedite the vehicle's destruction and render fire control or extinguishing efforts more challenging [2.18].

2.17.
Hot Surfaces

The regular operation of vehicles with a combustion engine substantially elevates the temperature of exhaust manifolds, catalytic converters, turbochargers, and exhaust gas recirculation (EGR) valves. These specific parts work at temperatures beyond the autoignition temperature of most automotive fluids. Therefore, extra precautions are taken during the design and manufacturing activities to avoid direct contact and mitigate the associated heat transfer mechanisms along the vehicle lifespan.

Additionally, it is crucial to recognize the potential ignition risk posed by the eventual contact of these elevated-temperature surfaces with external flammable substances such as grass, dry leaves, and litter, particularly in the lower regions of the vehicle.

However, under certain situations, other components may heat up and surpass the autoignition temperatures or create sparks. Some examples are improper lubrication, brake lockups, chassis friction against the road during a collision, and electrical failures.

2.18.
Electric Anomalies

Some electric anomalies and mechanisms can directly serve as energy sources capable of

initiating combustion reactions. Additionally, some electric processes have the potential to degrade materials and components, leading to overheating, carbonization, electrolytic corrosion, and subsequent fires. Chapter 4 recognizes the existence of this group of mechanisms and delves deeper into their descriptions and workings.

2.19.
Fire Accelerants

Fire accelerants are substances used intentionally to start or add to a fire. They are used to increase fire intensity or spread for malicious purposes such as vandalism and insurance fraud. Common fire accelerants include gasoline, alcohol, and paint solvents. The Analysis section of this book goes deeper into identifying arson by scrutinizing traces of accelerant usage.

2.20.
Heat Release Rate

The heat release rate (HRR), usually measured in kW, is a crucial parameter used for decades to measure and understand the behavior of several types of fires. It relates to peak temperatures, fire growth and spread, the likelihood of structural damages, difficulty in extinguishing, and firefighter safety. Since the HRR usually varies along the evolution of a fire, its peak value, PHRR, is often used as a more concise and meaningful indicator.

However, HRR and PHRR prove less informative when comparing ICEV, BEV, and HEV fires. Recent studies, like "A Review of Battery Fires in Electric Vehicles" by P. Sun et al. [2.21], show that measured values of PHRR are comparable among ICEV, BEV, and HEV of similar sizes and missions, as illustrated in **Figure 2.15**. However, fires with lithium traction batteries typically require much more time to put out. They also need ten times more water than fires in ICEVs.

Figure 2.15 PHRR comparison.

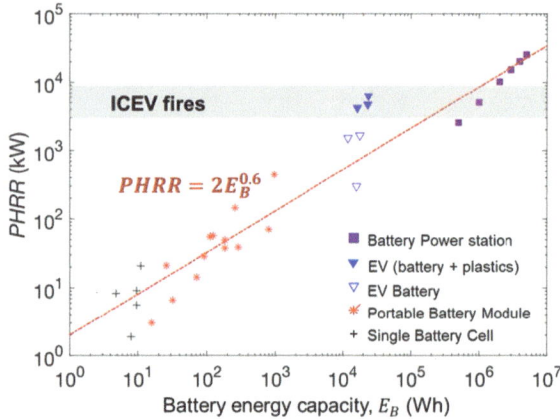

$$PHRR = 2E_B^{0.6}$$

Sun, P., Bisschop, R., Niu, H. *et al.* A Review of Battery Fires in Electric Vehicles. *Fire Technol* 56, 1361–1410 (2020), permission granted through Copyright Clearance Center courtesy of Springer.

2.21.
Fire Extinction Methods

The extinction of fire refers to ending the combustion reaction by eliminating or isolating one or more components of the fire tetrahedron. This entails isolating or removing the combustible substance, the oxidizer, and the energy source or disrupting the chain reaction. Various methods exist for extinguishing fires, and this book focuses on those particularly relevant to automotive fire scenarios. It is important to note that fire extinguishers are mandatory onboard automotive vehicles in certain countries, and

automated fire-extinguishing systems installed onboard are commonly found in race cars and large agricultural machinery.

Most motorcycles feature a straightforward safety mechanism: a safety valve, commonly known as a petcock, positioned beneath the fuel tank, as highlighted in **Figure 2.16**. This valve is designed for easy handling in emergencies, allowing a swift cessation of fuel flow and thereby preventing the fuel supply to the fire. Safety valves with the same intent are also used in vehicles powered by compressed natural gas (CNG) and LPG.

Figure 2.16 Bike Petcock valve.

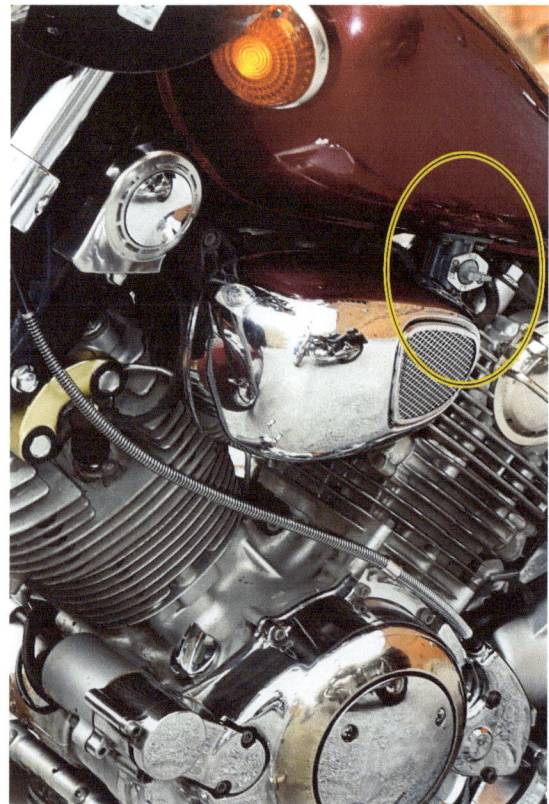

terekhovigor/Shutterstock.com.

2.21.1.
Water-Based Methods

Water is the most common and widely used extinguishing agent for fires. Water is readily available, inexpensive, and usually effective in reducing the temperature and depriving the fire of oxygen. However, water-based methods have some limitations and drawbacks. First, water is ineffective in some types of fires, such as electrical or metal. Moreover, under certain circumstances, water can pose a risk of electrocution if applied to live/exposed high-voltage circuits.

2.21.2.
Foam-Based Methods

Foam is a chemical fire-extinguishing agent commonly used to extinguish fires involving flammable liquids, such as petrol and oil. The foam creates a thick barrier between the fuel and the oxygen, which helps reduce the temperature and suffocates the fire. Foam is ineffective for electrical fires, and its use can damage electrical equipment or cause electrocution if applied to exposed high-voltage circuits.

2.21.3.
Dry Powder

Dry powder is a fire-extinguishing agent consisting of a fine powder of inert solid chemicals such as ammonium phosphate, sodium bicarbonate, and potassium bicarbonate. Dry powder forms a layer of powder over the fuel, interrupting the oxygen supply to the fire. It is safe for electrical fires since it does not conduct electricity. However, it leaves a residue that is difficult to clean and may damage sensitive electronic equipment.

2.21.4.
Inert Gases

Inert gases are gases that do not react with other substances and are, therefore, safe for use as fire-extinguishing agents. Some common inert gases used in fire extinguishers include argon, nitrogen, and carbon dioxide (CO_2). Inert gases displace the oxygen needed for combustion, thereby suffocating the fire. Inert gases are safe to use with sensitive electronic equipment since they do not leave any residue. They should not be used in confined spaces since they might cause human asphyxiation.

2.21.5.
Chain Reaction Interruption

Halon (a common denomination used for a series of halogenated gases, i.e., those containing bromine, chlorine, fluorine, and iodine) was used for several years as an effective fire-extinguishing agent. It interrupts the chain reaction by sequestering the crucial free radicals, halting the fire at much lower concentrations than required with other fire suppressants. Unfortunately, halon has been banned since it contributes to the depletion of Earth's ozone layer. Other gases that use the same principle without affecting the ozone layer were developed, such as FM-200 (heptafluoro-propane), and are being increasingly applied as a replacement. Since they are not conductive and leave no residues, they can be used in electronic equipment.

2.21.6.
Fire Blankets

Car fire blankets are meant to wrap the burning vehicle with a large, fire-resistant cloak, avoiding the entrance of oxygen, as seen in **Figure 2.17**. For ICEV fires (and xEV fires not involving the traction battery), they can contain the flames in a few seconds and put out the fire in a few minutes [2.19, 2.20]. Although they are ineffective in eliminating traction battery fires, they can aid in reducing the flames around an xEV fire and slow down the evolution of this type of fire.

Figure 2.17 Fire blanket.

Courtesy of Bridgehill.

2.21.7.
xEV Fire Extinction

Firefighting activities pose specific risks and challenges due to the elevated voltages in xEVs, while fires involving the traction battery are especially problematic. Chapter 8 will delve into the distinct difficulties and examine technologies to mitigate these issues [2.21].

References

[2.1]. Drysdale, D., *An Introduction to Fire Dynamics*, 3rd ed. (New York: John Wiley & Sons, 2011), doi:https://doi.org/10.1002/9781119975465.

[2.2]. National Fire Protection Association, "Guide for Fire and Explosion Investigations—NFPA 921," 2021 edition, 2021.

[2.3]. Quintiere, J., *Fundamentals of Fire Phenomena* (West Sussex, UK: John Wiley & Sons, 2006), doi:https://doi.org/10.1002/0470091150.

[2.4]. ASTM International Standard, "Standard Specification for Denatured Fuel Ethanol for Blending with Gasolines for Use as Automotive Spark-Ignition Engine Fuel," ASTM Standard D4806-21a, Revised October 2021.

[2.5]. International Organization for Standardization, "Petroleum and Related Products—Determination of Flash and Fire Points—Cleveland Open Cup Method," Standard ISO 2592:2017 (EN), 2017.

[2.6]. National Fire Protection Association, "Guide to Fire Hazard Properties of Flammable Liquids, Gases, and Volatile Solids—NFPA 325," 1994 edition, 1994.

[2.7]. Shields, L. and Scheibe, R., "Computer-Based Training in Vehicle Fire Investigation Part 2: Fuel Sources and Burn Patterns," SAE Technical Paper 2006-01-0548 (2006), doi:https://doi.org/10.4271/2006-01-0548.

[2.8]. Arndt, S., Stevens, D., and Arndt, M., "The Motor Vehicle in the Post-Crash Environment, An Understanding of Ignition Properties of Spilled Fuels," SAE Technical Paper 1999-01-0086 (1999), doi:https://doi.org/10.4271/1999-01-0086.

[2.9]. Shields, L. and Scheibe, R., "Computer-Based Training in Vehicle Fire Investigation Part 1: Ignition Sources," SAE Technical Paper 2006-01-0547 (2006), doi:https://doi.org/10.4271/2006-01-0547.

[2.10]. Babrauskas, V., "Ignition of Gases, Vapors, and Liquids by Hot Surfaces," in *Proceedings of the 3rd International Symposium on Fire Investigation Science and Technology (ISFI)*, College Park, MD, May 2008.

[2.11]. Byers, K., Epling, W., Cheuk, F., Kheireldin, M. et al., "Evaluation of Automobile Fluid Ignition on Hot Surfaces," SAE Technical Paper 2007-01-1394 (2007), doi:https://doi.org/10.4271/2007-01-1394.

[2.12]. Colwell, J.D. and Reza, A., "Hot Surface Ignition of Automotive and Aviation Fluids," *Fire Technology* 41 (2005): 105-123, doi:https://doi.org/10.1007/s10694-005-6388-6.

[2.13]. Somandepalli, V., Kelly, S., and Davis, S., "Hot Surface Ignition of Ethanol-Blended Fuels and Biodiesel," SAE Technical Paper 2008-01-0402 (2008), doi:https://doi.org/10.4271/2008-01-0402.

[2.14]. Federal Motor Vehicle Safety Standards, "Flammability of Interior Materials," FMVSS 302, Code of Federal Regulations, Federal Motor Vehicle Safety Standards, 49 CFR 571.302 Standard No. 302, October 2021.

[2.15]. International Organization for Standardization, "Road Vehicles, and Tractors and Machinery for Agriculture and Forestry—Determination of Burning Behaviour of Interior Materials," Standard ISO 3795, 1989.

[2.16]. SAE International Surface Vehicle Standard, "Flammability of Polymeric Interior Materials—Horizontal Test Method," SAE Standard J369, Revised August 2019.

[2.17]. Underwriters Laboratories Inc., "Standard for Safety Tests for Flammability of Plastic Materials for Parts in Devices and Appliances," UL-94, February 2023.

[2.18]. De Santis, T., Adams, C., Molnar, L., Washington, J. et al., "Motor Vehicle Fire Investigation," SAE Technical Paper 2008-01-0555 (2008), doi:https://doi.org/10.4271/2008-01-0555.

[2.19]. Bridgehill, "Bridgehill Car Fire Blanket," accessed May 2023, https://bridgehill.com/fire-blankets/car/.

[2.20]. Padtex Insulation, "Padtex Car Fire Blanket," accessed May 2023, https://padtex.lv/car-fire-blanket/.

[2.21]. Sun, P., Bisschop, R., Niu, H. et al., "A Review of Battery Fires in Electric Vehicles," *Fire Technology* 56 (2020): 1361-1410, doi:https://doi.org/10.1007/s10694-019-00944-3.

Chapter 03

Mechanical Sources of Heat, Stress, and Degradation

3.1.
Introduction

Mechanical degradations often result in two primary outcomes: functional failures and inadequate performance. However, specific instances of mechanical degradation can lead to vehicle fires through factors such as heat generation, the supply of fuels, and insufficient containment or control measures.

Cars are intricate machines composed of numerous interconnected mechanical systems, each susceptible to different forms of stress and degradation as time passes. Despite significant advancements in automotive engineering that have greatly improved vehicle safety, the possibility of fires still exists due to the complex interplay between mechanical components, their operational conditions, their wear out, and potential failure modes. This chapter examines the sources of mechanical deterioration that can ultimately evolve into car fires.

3.2.
Vibration

Vibration in an automobile refers to the oscillatory motion or shaking of various components, typically caused by the engine operation, road conditions, or other mechanical factors.

While some amount of vibration is expected in a running vehicle, excessive, uncontrolled, or prolonged vibrations can damage various components. In some cases, these failures may lead to fires [3.1].

3.2.1.
Sources of Vibration

3.2.1.1.
Engine

The primary source of vibration in an automobile is the engine. As the pistons move up and down, they generate reciprocating forces that create vibrations. Smaller parts like valves and oil pump gears also generate vibrations while the engine runs. Engine mounts isolate the motor from the chassis and absorb some vibrations, reducing the energy transfer through the mounting points. This enhances overall driving comfort and reduces the transmission of disruptive forces to the rest of the vehicle.

3.2.1.2.
Drivetrain

Vibrations can also arise from the transmission, driveshaft, and differential as power is transmitted from the engine to the wheels. These vibrations typically occur due to imbalances, asymmetries, misalignments, or worn-out components within these systems.

3.2.1.3.
Imbalanced Rotating Components

Several rotating parts like the flywheel, wheels, and tires cause vibrations if they are imbalanced, especially at higher rotational speeds. Imbalance occurs when uneven mass distribution around the rotating axis leads to asymmetric centrifugal forces. These forces induce oscillatory efforts on the stationary parts. Proper wheel balancing, captured in **Figure 3.1**, can at least reduce some of these stresses.

Figure 3.1 Wheel balancing.

Sergey Ryzhov/Shutterstock.com.

3.2.1.4.
Road Conditions

Uneven road surfaces, potholes, and bumps induce vibrations in the suspension system. While this system mitigates and minimizes these vibrations, it cannot eliminate all the associated energy. Moreover, worn-out or damaged components, like shock absorbers, bushings, or control arms, can exacerbate these vibrations.

3.2.2.
Damaging Effects

3.2.2.1.
Cracks, Fractures, and Fatigue Failure

Intense vibrations can cause significant damage to various components, particularly those composed of brittle materials like cast iron or ceramic. These fierce vibrations can result in cracks or fractures, jeopardizing the structural integrity of the affected parts. Furthermore, when vibrations persist over an extended period, they induce cyclic stress on components, which may trigger fatigue failure, as exemplified in **Figure 3.2**.

Figure 3.2 Fatigue fracture.

For instance, if an exhaust manifold cracks, an unexpected flow of hot gases may reach flammable parts underneath the vehicle, initiating combustion.

3.2.2.2.
Wear and Tear
Vibrations can accelerate the wear and tear of various parts by causing friction and rubbing between components, including bearings,

harnesses, pipelines, and seals. Even protection devices, like the corrugated hoses depicted in **Figure 3.3**, can be ruined due to excessive wear against a sharp edge. Consequently, over time, this will allow damage to an internal fuel pipeline or the plastic insulator of an electric wire these corrugated hoses were trying to protect.

Moreover, specific components endure a combination of mechanical stresses, such as brake hoses. During regular vehicle operation, these hoses are exposed to vibrations from the road and engine, bending stresses caused by suspension and steering movements, and internal pressure when the brake pedal is pressed. In **Figure 3.4**, laboratory results from a simulation conducted by Cantoni et al. demonstrate that the brake hoses tend to fracture near the neck of their metallic connection, a region of concentrated mechanical stresses [3.2]. Once the hose develops cracks, the flammable brake fluid can potentially spill beneath the vehicle, creating a significant fire hazard.

Figure 3.3 Wear of corrugated protection hoses.

Figure 3.4 Fatigue rupture of brake hoses.

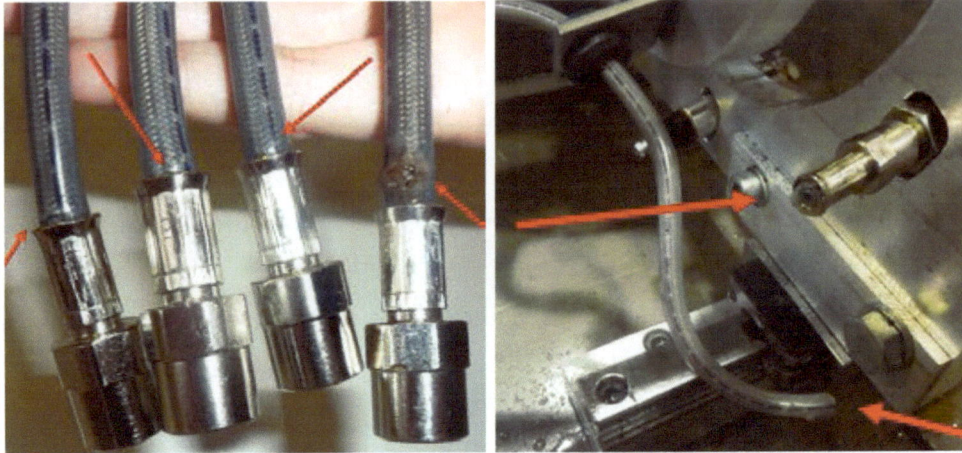

3.2.2.3.
Fasteners and Bolts

Vibrations can gradually loosen fasteners, including bolts and nuts, and unlock other types of fasteners, like clamps and clips, harming various systems. If adequate countermeasures are not in place, this phenomenon may lead to misalignments, detachments, and component failures over time. Additionally, misalignments can lead to increased wear and compromised sealings. Consider the occurrence of a misaligned casing of a hydraulic pump, allowing the leakage of its fluid to be sprayed over the catalytic converter. The fire risk becomes evident.

3.2.2.4.
Electronics and Wiring

Delicate electronic components and wiring harnesses are sensitive to excessive vibrations, potentially leading to malfunctions or connection issues. Examples of damage that can occur include the fracturing of solder joints, loosening of connections, and occurrence of fretting corrosion. Some of these failures and their relationship with automotive fires will be further examined in Chapter 4.

3.3.
Impacts

Mechanical impacts on automotive components, particularly those resulting from vehicle collisions, can have detrimental effects, such as fuel line ruptures and damage to electrical wiring, increasing the risk of fire. Furthermore, collisions can lead to contact between heated exhaust system parts and combustible materials within the vehicle, including plastics from connectors and hoses, further exacerbating the potential for fires [3.3-3.5].

Figure 3.5 shows that some of the wire harnesses were affected. If repairs are poorly conducted, they might pose an immediate fire hazard or become a risk later.

Other unexpected impacts may degrade automotive parts before the vehicle is delivered to the end customer or along the vehicle usage. Consider the case of an electronic module dropped on the factory floor before being installed and the case of a module fixed in a low place, subjected to eventual gravel hits. Even if an immediate failure is not observed, internal solder joints, for instance, might be weakened by these impacts and lead to failures [3.6].

Figure 3.5 Car damaged by collision.

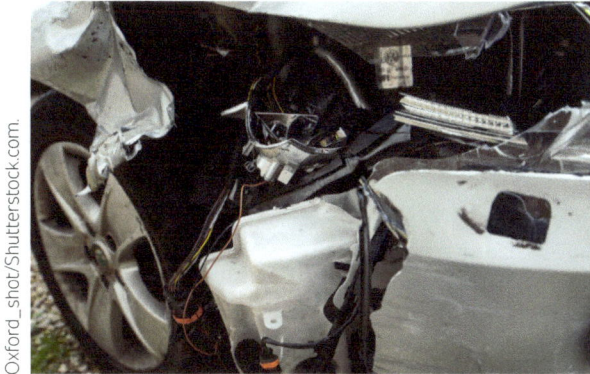

Oxford_shot/Shutterstock.com.

3.4.
Pressure

The vehicle's fuel system and other hydraulic systems apply controlled pressure on their fluids. However, excessive pressure poses significant risks to these systems, and their capability to endure normal pressure can degrade along the vehicle's life. Let us examine the implications in more detail.

3.4.1.
Fuel System

The fuel system is designed to deliver fuel to the engine in a managed manner, and modern vehicles often use high-pressure fuel injection. Excessive pressure in the fuel system can rupture fuel lines, fittings, injectors, or sensors and cause fuel to spray or leak out. Additionally, high pressure can cause sealings, such as O-rings, to fail and leak, further compromising the integrity of the fuel system. Fuel dispersion increases the fire risk, especially if it reaches a nearby ignition source.

3.4.2.
Hydraulic and Lubrication Systems

Many automotive systems, such as power steering, braking, and lubrication, rely on hydraulic

pressure. If there is excessive pressure in these systems, it can cause leaks or bursts in hoses, pipes, connections, or seals, leading to functional failures and potential fire hazards if their fluids contact hot engine components or electrical systems.

3.5.
Cuts

This type of damage may, for instance, rupture the envelope of a fuel line, allowing its fluid to reach nearby ignition sources. It might also be the case where the plastic insulator of an electric wire is severed, allowing a short circuit to ground. Or it could permit water entry and future damage to the attached electronic module, as will be seen in Chapter 6 in more detail.

Given the abundance of sharp edges in the vehicle's metallic parts, adequate caution and containment measures must be taken in vehicle design, manufacturing, and maintenance.

3.6.
Crushing

Crushing a wire or a pipeline often damages the plastic insulator or the hose envelope, with consequences like the ones caused by cuts. This damage is relatively more frequent to electric harnesses since stranded wires are more common and more difficult to see along maintenance and repair operations, particularly of large parts.

3.7.
Stretching

When a hose is stretched, it might become loose from its connection or rip apart. Similarly, when an electric wire is pulled, it might damage the connector at one of its ends or degrade its contact resistance, or the wire may break apart. Another possible consequence is the degradation of the electric connector sealing, leading to water entry and related damages.

3.8.
Friction

Friction in a vehicle can lead to fires due to heat generation and its effect on nearby materials. Friction converts mechanical energy into heat energy and, in some cases, significant wear of the rubbing parts. Usually, the heat generated from friction is relatively small and dissipated locally or managed by the vehicle's cooling systems. Still, under certain circumstances, excessive heat can build up and potentially ignite flammable materials. Abnormal conditions include locked brakes, insufficient lubrication, dysfunctional cooling systems, and fretting between vehicle parts and pavement during collisions.

In the case of large vehicles, such as buses and semitrucks, a brake failure occurring in one of the rear wheels can go unnoticed by the driver. Furthermore, in the case of semitrucks, there exists the possibility of "driving through" the parking brakes, meaning the vehicle can continue to move even with the parking brakes engaged. These situations are frequently ignored due to the engine's substantial power and torque reserve and the significant distance between the driver and the specific wheel or wheels affected [3.7].

As a result of these unnoticed issues, the brake drum's temperature can increase to the point where it heats the corresponding wheel to the extent that it ignites its tire, potentially leading to a fire. This concerning phenomenon is illustrated in **Figure 3.6**, sourced from a study conducted by Parrott, K., and Stahl, D [3.7].

Figure 3.6 Brake fire.

Reprinted from SAE Technical Paper 2013-01-0207.
© SAE International.

3.9.
Rodents

Rodents like mice and rats instinctively gnaw on various objects, including electrical wires and plastic fuel lines. They do this to control their teeth growth and mark their territory. Also, as a food source, soy-based and biodegradable materials in wiring harnesses might be attractive to these pests. When rodents infest a car, they may find their way into the engine compartment or other vehicle regions where electrical wiring or plastic hoses are present. They can chew through the protective insulation of the wires, exposing the conductive metal beneath, and even cut through the copper filaments, as illustrated in **Figure 3.7**. If the exposed wires or conductive components contact each other, this can create short circuits. The high current flow generates heat and sparks, potentially igniting nearby flammable materials. Also, if the bites puncture a hose, the subsequent fluid spill could initiate a fire.

Moreover, rodents often seek shelter in warm and secluded areas, such as the engine compartments of parked vehicles. They may carry flammable materials, such as dried leaves, twigs, or paper, into these spaces to build nests. If later on, these materials contact hot engine components or the exhaust system, they can catch fire and quickly spread to other parts of the vehicle.

Figure 3.7 Rodent damage.

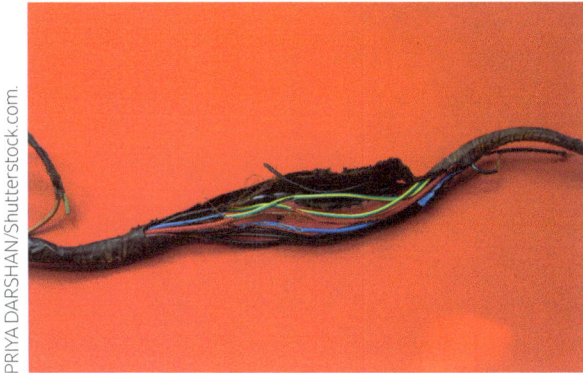

PRIYA DARSHAN/Shutterstock.com.

3.10.
Mechanical Overloads

Mechanical overloads can lead to car fires due to the generation of excessive heat and failure of

critical components. While some examples, such as the structural damage caused by cargo overload depicted in **Figure 3.8**, are easily noticeable, others may be more subtle or gradual.

For instance, prolonged uphill towing can lead to engine overheating, while continuous brakes downhill can make the brake pads or shoes and surrounding parts overheat.

Specific components like connecting rods, crankshafts, or pistons can experience excessive stress when subjected to continuous mechanical overloads. This prolonged stress can weaken the metal structure, leading to deformation, cracking, or even fracture. The compromised integrity of these components can ultimately lead to catastrophic failures. These situations may not be immediately apparent but can harm the vehicle's safety over time, including fire risks.

Figure 3.8 Cargo overload.

Kwanchanog Noinwong/Shutterstock.com.

3.11.
Degradation of Pipelines, Seals, and Hydraulic Connections

The degradation of the exhaust system, pipelines, seals, and fluid connections in a car can lead to car fires due to the release of hot gases, flammable substances, and volatile materials. A relatively frequent and often neglected situation is illustrated in **Figure 3.9**: leakage of lubrication oil along the engine–transmission interface of an old car.

Figure 3.9 Oil leakage on engine-transmission interface.

Courtesy of Erbis Llobet Biscarri.

Leakages in the cooling system, as exemplified in **Figure 3.10**, can reduce the volume of its fluid to a point where cooling becomes insufficient and the engine or transmission overheats.

Figure 3.10 Leaky coolant reservoir.

Oasishifi/Shutterstock.com.

Various hydraulic connections exist in a car's engine, transmission, and other systems. Over time, these can become loose or degrade, developing leaks. Consider a damaged connection in one of the engine's fuel lines captured in **Figure 3.11**: diesel can be sprayed over the motor, which can cause a fire.

Figure 3.11 Connections of a diesel pump.

Voyagerix/Shutterstock.com.

References

[3.1].   Pang, J., *Noise and Vibration Control in Automotive Bodies* (Beijing, China: China Machine Press, 2018), doi:https://doi.org/10.1002/9781119515500.

[3.2].   Cantoni, C., Mastinu, G., Gobbi, M., Ballo, F. et al., "Accelerated Testing of Brake Hoses for Durability Assessment," *SAE Int. J. Commer. Veh.* 10, no. 1 (2017): 178-183, doi:https://doi.org/10.4271/2017-01-0389.

[3.3].   Brach, M., Brach, R., and Mason, J., *Vehicle Accident Analysis and Reconstruction Methods*, 3rd ed., SAE International Book R-516 (Warrendale: SAE International, 2022), ISBN:978-1-4686-0419-1.

[3.4].   Struble, D. and Struble, J., *Automotive Accident Reconstruction: Practices and Principles* (Boca Raton: CRC Press, 2020), doi:https://doi.org/10.1201/9781003008972.

[3.5].   Hande, M. and Wojcik, M., "Failure Assessment of Electronics Component under Gravel Bombardment Test in Automotive Electronics," SAE Technical Paper 2023-01-0159 (2023), doi:https://doi.org/10.4271/2023-01-0159.

[3.6].   Hande, M. and Kumar, V., "Failure Prediction & Mitigation of Electronic Component in Automotive Controller during Transient Dynamic Event," SAE Technical Paper 2017-26-0356 (2017), doi:https://doi.org/10.4271/2017-26-0356.

[3.7].   Parrott, K. and Stahl, D., "Case Studies of Parking Brake Fires in Commercial Vehicles," SAE Technical Paper 2013-01-0207 (2013), doi:https://doi.org/10.4271/2013-01-0207.

Chapter 04

Electric Sources of Heat and Degradation

4.1.
Introduction

Numerous components work in harmony within the intricate network of electrical systems, providing enhanced functionality, comfort, convenience, and safety features. However, when subjected to excessive stress and on the passage of time, these systems may experience failures that can escalate into abnormal heat generation and trigger car fires.

One of the primary sources of stress in an electrical system is the degradation of connections. Over time, corrosion, loose connections, and poor wiring practices can increase resistance and generate excessive heat. Similarly, electric leakages, harness degradations, and improper installation of accessories can all contribute to the risk of electrical faults and subsequent fires.

Furthermore, the realm of electric transients presents various challenges for electronic systems. Events such as cranking, load dump, inductive spikes, reverse battery, and electro-static discharge (ESD) can impose overwhelming stress on the electrical system, potentially destroying sensitive electronic components and igniting fires.

Another critical concern is short circuits, which can arise from damaged insulation, overloads, or faulty connectors. A sudden current surge through an unintended pathway can generate intense heat and sparks, escalating into a full-blown fire.

Occasionally, natural phenomena like lightning strikes can directly trigger car fires by overwhelming the vehicle's electrical system with a massive surge of energy. Additionally, electronic components that fail due to manufacturing defects, age-related deterioration, or random failures can become ignition points if they generate heat or sparks near flammable materials [4.1].

4.2.
Intended/Intrinsic Heat Generators

Several vehicle components, such as filament lamps, diesel glow plugs, defrosters, HVAC systems, and seat heaters, are designed to generate heat as part of their regular operation. However, in cases of component failures or control module malfunctions, these temperatures can exceed safe limits, posing a fire hazard. Moreover, during collisions, the glass envelope of a light bulb (illustrated in **Figure 4.1**) can shatter, exposing the hot filament, which can reach temperatures around 4500°F (2500°C). This scenario can lead to contact with flammable materials or vapors. While recent designs employ LED light sources at much lower temperatures (below 210°F or 100°C), many vehicles on worldwide streets still use filament lamps. Therefore, understanding the potential relationship between filament lamps and car fires remains important.

Figure 4.1 Filament lamp.

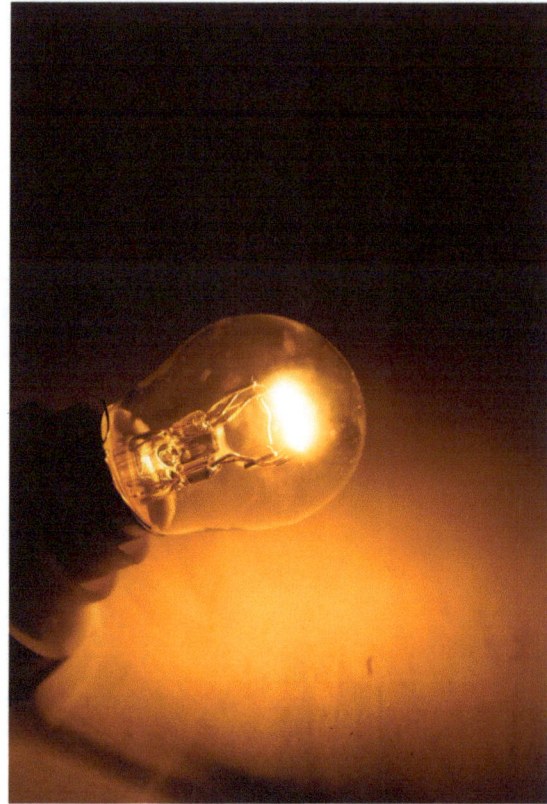

Marco Tjoanda/Shutterstock.com.

4.3.
Electric Overload/Series Resistance Increase

Electric overloads and increased series resistance can lead to fires in electrical systems. Both mechanisms can elevate the temperature of wires or metallic terminals and, consequently, the surrounding plastic insulators. Eventually, these polymers melt or degrade to a point where they allow metal-to-metal contact, causing short circuits. Additionally, under higher temperatures, the insulators may undergo carbonization, creating undesired leakage currents and exacerbating the heating process. Let us discuss each of these factors separately.

4.3.1.
Electric Overload

When the amount of current flowing in a circuit surpasses its capability to handle that amount safely, an electric overload happens. This can occur when a faulty module draws more current than the circuit can safely provide or when too many loads are fed from a single supply line. Given that electrical conductors, such as wires and cables, and conductor components in connectors and switches have an equivalent resistance, the Joule effect implies that heat will be dissipated in a quadratic relationship to the circulating current. In other words, a slight increase in the circulating current will mean a quadratic increase in the generated heat. If the components cannot dissipate the excessive heat, temperatures will rise substantially. **Figure 4.2** displays an example where the fuse receptacle deteriorated before the fuse opened; the consequences could have been much worse.

Figure 4.2 Melted fuse receptacle.

Serrgey75/Shutterstock.com.

4.3.2.
Series Resistance Increase

Consider the simple circuit of **Figure 4.3**, which consists of an electric motor powered by the vehicle battery through a harness and a connector. The lower part of the figure illustrates the equivalent circuit, providing a more detailed representation of

the critical components. Under normal circumstances, the series contact resistances of the connector terminals, R1 and R2, are negligible compared to the load resistance RL. In this situation, the heat dissipated in R1 and R2 does not cause a significant temperature rise in these connections.

However, let us imagine a scenario where R1 experiences a significant rise in resistance due to a loose or oxidized connection [4.2]. Because of the Joule effect, the heat generated at this point will intensify. If this increase becomes excessive, it can result in an increase in temperature to the extent that nearby insulation materials melt or carbonize. Melted insulators may allow direct contact with conductors; therefore, short circuits may occur, while carbonized parts will enable leakage currents, both effects creating fire hazards.

Figure 4.3 Series resistance increase.

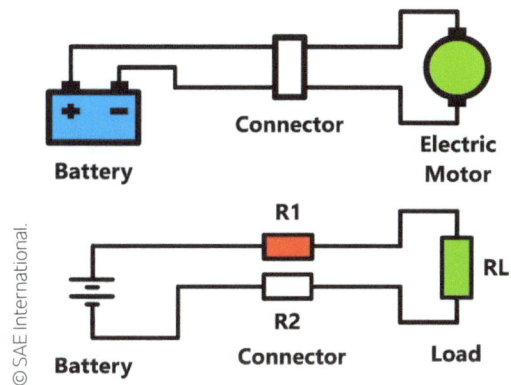

© SAE International.

4.4.
Electric Leakages/Parallel Resistance Decrease

In an automotive context, electric leakages refer to the unplanned flow of electrical current through unintended paths or circuits, often caused by damaged or improperly installed electrical components, wiring faults, or insulation failures. They can also result from water ingress and may lead to short circuits.

From an equivalent circuit perspective, these situations imply a leakage resistance parallel to the original load, as illustrated in **Figure 4.4**. R1 represents a leakage between one of the power wires and the vehicle chassis, while R2 represents a leakage inside the electric motor. Compared to the electric overload scenario, the main concern here regarding fire risks is the heat generated along the leaky path. This heat can significantly increase the temperature of nearby materials, leading to their meltdown or carbonization.

Figure 4.4 Electric leakages.

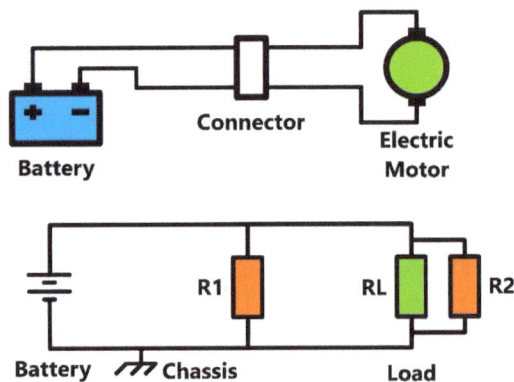

© SAE International.

Some items in this chapter will provide more examples of failure mechanisms associated with water. In contrast, Chapter 6 will examine how water can ingress sensitive electronic modules and parts in more detail.

4.5.
Harness Degradations

The wiring harness is a complex network of electrical cables connecting various vehicle

modules and devices to enable their proper energization, communication, and functioning [4.3, 4.4]. Each electric wire, known as automotive wire, typically consists of a copper core encased in a protective plastic covering. The cross-sectional area of copper is calculated to minimize resistance within each electric circuit, among other premises. To enhance flexibility, automotive wires are composed of multiple fine wire strands. This flexibility is crucial during manufacturing, allowing operators to route the harnesses along the vehicle structure and manipulate their ends to attach various connectors. Additionally, flexibility is vital in harnesses subjected to numerous bending cycles throughout the vehicle's lifespan, such as door connecting harnesses. The plastic covering provides electrical insulation to prevent short circuits and acts as a mechanical barrier, shielding the copper from contact with water and other fluids.

Chapters 3 and 5 discuss degradation mechanisms affecting several components, including harnesses. In this chapter the primary emphasis is on electric damages that can potentially result in car fires. The items in the following sections present an organized categorization of harness degradations, accompanied by a list of events that can induce such deteriorations.

4.5.1.
Series Resistance Increase

When the effective cross-section area of copper diminishes, the series resistance of a cable will increase. Some events that can cause this situation are exposure of the copper core to aggressive chemicals, particularly sulfuric acid from the 12 V lead battery. Additionally, the series resistance can be increased by the rupture of individual copper filaments due to excessive stretching, fatigue, cuts, abrasion, or rodent bites. Suppose the current flowing through a wire with this type of damage is high enough. In that case, it will overheat and eventually escalate to the ignition of surrounding materials or cause the complete rupture of the wire.

4.5.2.
Cable Rupture

In this case, the reference event is the complete rupture of the cable, interrupting the current flow. This might create significant arcing during the breakup of the wire, especially at high currents, high voltages, or inductive loads. The electric arc will elevate the temperatures of the surrounding insulators that, in turn, might catch fire. Another possibility is the intermittent contact between the two copper ends around the rupture point, leading to several sparks.

Excessive traction forces, fatigue, cuts, abrasion, rodent bites, and excessive current (overloads) are events capable of breaking electric cables. In the case of excessive current, the cable will achieve high temperatures and damage the plastic insulators before the copper melts, which, in turn, might create short circuits and leakage currents.

It is worth noting that plastic insulators commonly used in automotive harnesses have a melting point of around 220°F (105°C), while copper melts at a much higher temperature of 1984°F (1085°C). Therefore, the plastic insulator may deform well before the copper reaches its melting point and opens the circuit in situations involving excessive current. This can further contribute to the occurrence of short circuits. Also, if the insulators carbonize, leakage currents will escalate.

4.5.3.
Short Circuits

A short circuit occurs when an unintended connection or low-resistance pathway is formed between two points in an electrical circuit. This bypasses the normal resistance or load in the circuit, allowing a significant amount of current to flow through the shorted path. The situation depicted in Figure 4.5, commonly called a *dead* short circuit, implies a violent temperature increase and spark generation.

Figure 4.5 Dead short circuit.

Porntep Naprasert/Shutterstock.com.

However, less dramatic events result in the overheating of materials surrounding the abnormal current path without generating sparks, known as *limited* short circuits. One example is the current flowing through a carbonized insulator. From a fire risk perspective, dead and limited short circuits can potentially escalate into a car fire.

4.5.4.
Copper Beads

Copper beads, encountered in car fire investigations, are small, irregularly shaped copper fusion remnants found within a vehicle's wiring harness and surrounding components, serving as relevant evidence for deciphering the fire's origin and progression. Notably, typical vehicle fire temperatures remain below the melting point of copper of approximately 1984°F (1085°C), necessitating an additional heat source such as short circuits or arcing for copper bead formation. Short circuits generate intense localized heat when an abrupt surge of electrical current flows through a wire, leading to partial copper melt and bead formation, which can either remain attached or detach and drip.

The presence of copper beads within a vehicle's wiring harness can suggest that a short circuit may have triggered the ignition of fire. Nevertheless, it is crucial to acknowledge that copper beads can also result from secondary damage within a car fire originating from different causes. In vehicles, various wires remain energized even during idling, and a fire in one area may lead to the melting or burning of plastic wire insulation. Consequently, the exposed copper core can contact other wires or the metallic chassis, causing short circuits and potentially forming copper beads.

Hence, while copper beads offer valuable insights into car fires, investigators must thoroughly examine the broader context and gather additional evidence to establish the true origin of fire and sequence of events. Historically, the presence of copper beads was often considered conclusive evidence of a short circuit as the root cause. However, recent studies and experiments have demonstrated that copper beads may also manifest as secondary occurrences in fires initiated by thermal/chemical factors, such as gasoline leakage. **Figures 4.6** and **4.7**, reproduced from a study by DeMarois et al. [4.5], illustrate instances of copper beads in vehicles subjected to deliberate chemical fires initiated in the driver's seat.

Figure 4.6 Copper beads along a wire.

Reprinted from SAE Technical Paper 2018-01-1439. © SAE International.

Figure 4.7 Copper beads on wire ends.

Reprinted from SAE Technical Paper 2018-01-1439. © SAE International.

4.5.5.
Water Intrusion

Automotive wires are designed and manufactured impervious to water, to ensure reliable electrical connections and prevent water-related damage. However, physical damages can compromise their watertightness. When the plastic insulator is cut, perforated, or cracked, water can enter through these openings and travel along the wires, eventually reaching connectors and sensitive electronics.

Various factors can contribute to physical damage to sealings, including collisions, abrasion, exposure to extreme temperatures, mishandling during repair procedures, or the use of aggressive diagnostic tools. For instance, the improper use of needle probes, as illustrated in **Figure 4.8**, can harm the sealing of connector terminals. Another detrimental practice involves perforating the wire insulator with such probes to establish contact with the copper wire.

Figure 4.8 Inadequate use of needle probes.

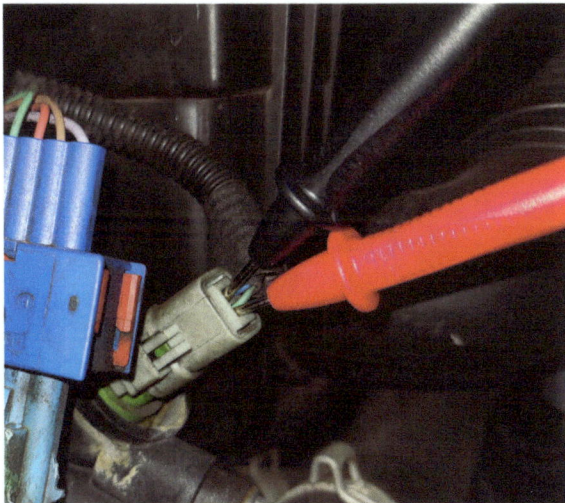

Courtesy of Erbis Llobet Biscarri.

Furthermore, the passage of time can take its toll on materials used for wire insulation. Environmental influences, such as prolonged exposure to heat, sunlight (and the associated UV radiation), and certain chemicals, can contribute to the gradual degradation of insulation materials. As this degradation progresses, the insulation may contract or develop cracks, compromising its watertight properties.

Chapter 6 will examine how capillarity and "respiration" can cause water to travel long distances inside the harness and reach far-away connectors and electronic modules.

4.5.6.
Leakage Current

Water intrusion in automotive wires can lead to various outcomes. One such outcome is the generation of a leakage current when water encounters the inner copper core and surrounding metallic parts (typically connected to the battery ground) or other exposed wires simultaneously. In certain situations, this leakage current will cause functional failures, but it might also evolve into electrolytic corrosion, local heating, deterioration of nearby insulators, and, ultimately, vehicle fires.

Additionally, suppose the leakage current is a consequence of carbonized insulators. In that case, the heat generated by this current will likely further carbonize the surrounding materials, escalating the current value and the temperature. This escalation in current and temperature significantly increases the risk of a fire outbreak.

4.6.
Connection Degradations

Automobile harnesses require connectors to establish reliable electrical connections, simplify installation and maintenance, provide flexibility and modularity, withstand harsh operating conditions, ensure compatibility, and facilitate future upgrades and modifications. Typically, a connector consists of a plastic case, multiple terminals, and sealing and locking features [4.6].

Unfortunately, they are subject to several mechanical, chemical, and electrical failure mechanisms. **Figure 4.9** shows an example: the blue plastic housing overheated and melted, while some connecting wires also exhibit signs of overheating. Some connector failures can result in functional anomalies and, in extreme cases, ignite a fire.

Figure 4.9 Damaged connector.

Pelagija/Shutterstock.com.

4.6.1.
Loose Ground Connection

A typical vehicle establishes the electrical ground connection, in low-voltage systems, by connecting components, such as lamps, electric motors, and electronic modules, to the vehicle's chassis or metallic body, as exemplified by **Figure 4.10**. Since the battery negative pole is also connected to the chassis or body, electric current flows from the components back to the battery, completing the circuit without needing individual ground wires for each load.

Figure 4.10 Typical ground connection in the engine compartment.

kaninw/Shutterstock.com.

However, intermittent connections and arcing can arise if the ground connections are loose, oxidized, or contaminated. Additionally, these faulty connections increase the series resistance, resulting in malfunctions, overheating, and, in worst cases, car fires. Modi and Galgoci's laboratory investigation demonstrated that even with modest currents, approximately 50 A, a loosely bolted wire connection could reach temperatures soaring into the several hundred degrees Celsius range [4.2].

4.6.2.
Incomplete Connector Mating

Most automotive connectors are equipped with positive locking mechanisms to ensure secure mating. These mechanisms are designed to withstand normal vibration and traction forces, preventing accidental disengagement of the connector. However, perfection is elusive. Factors such as incorrect assembly, debris accumulation, deformation, or fracture of the locking mechanism can compromise the proper mating of the connectors. Over time, these issues can lead to increased series resistance, intermittencies, or open circuits. These, in turn, may cause functional failures, arcing, overheating, and, in extreme cases, car fires.

4.6.3.
Deformed and Broken Contacts

Male and female terminals establish secure electrical connections between mating connectors. These terminal types are distinguished by their physical characteristics and how they interlock.

- Male Terminal: A male terminal is characterized by a protruding pin or plug-like structure designed to fit into a corresponding female terminal. When inserted into the female terminal, it typically has a solid metal pin or blade that provides electrical contact.

- Female Terminal: A female terminal is designed to receive and make electrical contact with a male terminal. It typically has a socket or receptacle with a hole, slot, or chamber where the male terminal can be inserted. The female terminal is usually designed to securely hold the male terminal in place and establish a reliable electrical connection.

Also, different approaches are employed to ensure low contact resistance. In some cases, a spring is utilized to exert pressure between the terminals, while in other cases, the geometry and base material of the terminals provide the necessary spring effect. For instance, a U-shaped female terminal made of beryllium copper alloy can effectively serve this purpose.

However, deformation or breakage of individual contacts and damage to the springs can result in increased contact resistance, intermittent connections, and open circuits. The consequences of these issues are like those discussed in previous topics. Common causes include wear and tear, excessive engagement/disengagement operations, vibration fatigue, overheating, and manufacturing defects.

4.6.4.
Pushed Back Pins

Particularly in compact, multiple-pin connectors, the minuscule locks responsible for securing each terminal are prone to damage. When these locks are impaired, there is a risk of the corresponding contact being dislodged along the connector engagement without the operator's awareness. As a result, this compromised electrical connection can give rise to the same issues outlined in the preceding points.

4.6.5.
Bad Crimp

The crimp joint is the most common connection between an automotive wire and a connector's

contact. **Figure 4.11** depicts an example of a crimp type terminal supplied by reel. The automotive wire, already stripped, has its copper filaments placed inside the U-section of a virgin terminal. A special press crimps the U, ensuring adequate pressure on the filaments and an adequate profile for the resulting envelope. At the same time, the two uppermost arms of the terminal are bent, ensuring that the plastic insulator will not move. The terminal is also detached from the strip, resulting in the terminal being firmly attached to the wire, as shown on the right end of the photo.

Figure 4.11 Crimped terminal example.

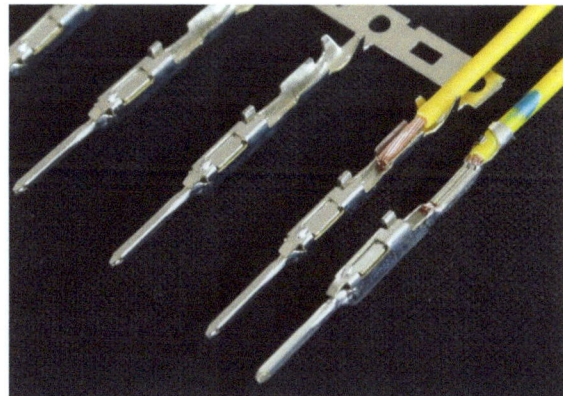

Reprinted from SAE Technical Paper 2008-01-1271.
© SAE International.

Possible anomalies include insufficient crimp force, which can lead to a loose connection, and excessive crimp force. In excessive force, the copper filaments within the wire may break or the terminal envelope can be fractured, leading to a compromised and loose connection [4.7].

4.6.6.
Oxidation and Corrosion

Oxidation and corrosion are related phenomena but not the same, although they often co-occur in the context of electric contacts [4.8, 4.9].

Oxidation refers to a chemical reaction in which a substance combines with oxygen. When metals are exposed to air, they can undergo oxidation, forming a layer of oxide on their surface. This oxide layer can act as a barrier, preventing further oxidation and protecting the underlying metal. In some cases, the oxide layer is a thin film that does not degrade the part's conductivity. However, in other cases, it may increase substantially the resistance and affect the performance of electric contacts.

Corrosion is a broader term that encompasses various chemical reactions between a material (usually a metal) and its environment, resulting in the degradation of the material. Corrosion can involve oxidation as one of the reactions, but it can also include other processes such as reduction, dissolution, or electrochemical reactions. In the case of electric contacts, corrosion can occur due to exposure to moisture, salts, acids, or other corrosive substances in the surrounding environment. Corrosion can cause physical damage, degradation, or alteration of the contact surfaces, leading to increased resistance, intermittent contact, or complete failure.

When exposed to ambient air and humidity, most copper alloys are prone to oxidation and corrosion. In electric applications, copper-based terminals and contacts are often plated with metals less susceptible to these degradations, such as nickel or gold. However, oxidation and corrosion reactions can still occur with a plating layer containing pores or cracks, as exemplified in **Figure 4.12** cross-section view. As a result of these reactions, corrosion by-products form and occupy a larger volume, eventually reaching the nickel surface and spreading around the aperture. Contact resistance can significantly increase if these by-products become trapped in the contact area.

Figure 4.12 Copper corrosion through platting pore.

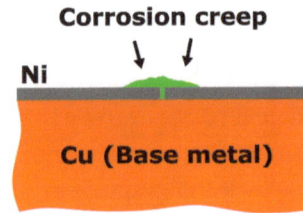

© SAE International.

4.6.7.
Fretting Corrosion

This degradation mechanism occurs when small movements take place between two contacting surfaces. These movements, typically between 0.04 and 4 mils (approximately 0.001 to 0.1 mm), can be triggered by vibration or thermal expansion and contraction cycles. Over time, after a few thousand cycles, the oxide accumulation impedes direct contact between the two terminals. This oxide buildup increases series resistance, leading to functional failures or local overheating [4.10].

To better understand the progression of fretting corrosion, let us refer to the cross sections around the point of contact of two terminals depicted in **Figure 4.13**.

Figure 4.13 Fretting corrosion.

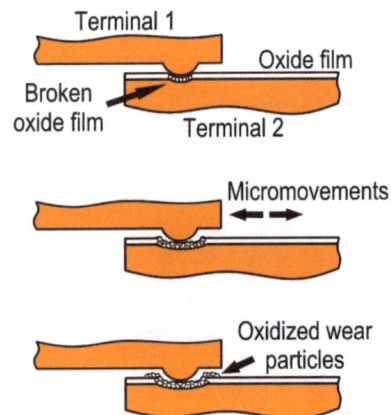

© SAE International.

The upper portion of the illustration captures the initial scenario immediately after the connection is made. Assuming that Terminal 2's contact surface is made of tin, there is already a skinny layer of tin oxide. As Terminal 1 comes into contact, it slightly deforms and cracks this oxide layer. The resulting increase in the series resistance in the contact area is negligible at this stage.

As the process continues and several micro-movements or fretting cycles occur, the brittle oxide layer experiences repeated fracturing, as captured in the middle of the drawing. This fracturing gives rise to hard grains and scales that dig into the metal, forming new grains as they are exposed to the surrounding air. This exposure leads to further oxidation of these newly formed grains.

Finally, the lower portion of the illustration exhibits the situation after several thousand cycles. At this stage, the contact region of Terminal 2 becomes grooved, and a thick layer of oxidized wear particles is dispersed around it. This accumulation of oxidized particles substantially increases the contact resistance, often by around two orders of magnitude.

4.6.8.
Leakage Currents

Leakage currents in connectors can arise from the carbonization of the plastic housing due to exposure to high temperatures, the intrusion of water, or contaminants. These phenomena create unintended conductive paths between terminals, leading to functional failures, potential damage to the connector terminals and surrounding components, temperature increases, and safety hazards.

4.6.9.
Damaged Housing

Most automotive connectors come equipped with plastic housings, which have limited ruggedness. These plastic housings can become damaged under various conditions, such as excessive mechanical stresses, mishandling, extreme temperatures, prolonged exposure to IR and UV radiations, aggressive chemicals, and normal wear and tear. If the housing is compromised, it can lead to several undesirable consequences, including loss of sealing, intermittent connections, arcing, functional failures, leakage currents, overheating, and even short circuits.

4.7.
Electric Transients and EMI

The ICEV electric system, commonly known as the "12 V" system, experiences various voltage fluctuations in different situations [4.11-4.14]. **Figure 4.14** illustrates some of these events. Usually, when the engine is off and there are no loads, a fully charged lead–acid battery exhibits approximately 12 V at ambient temperature. However, an almost depleted battery can show a steady-state voltage of around 10 V. Conversely, under extreme conditions, such as a high-speed engine running with a charged battery in cold weather, the voltage can reach 16 V in a steady state.

Furthermore, the 12 V system is subjected to numerous transient events. During engine cranking, particularly in cold weather, the voltage can momentarily plummet to around 6 V for hundreds of milliseconds before gradually recovering its standard value. While this under-voltage condition generally does not cause

Figure 4.14 Typical electric transients.

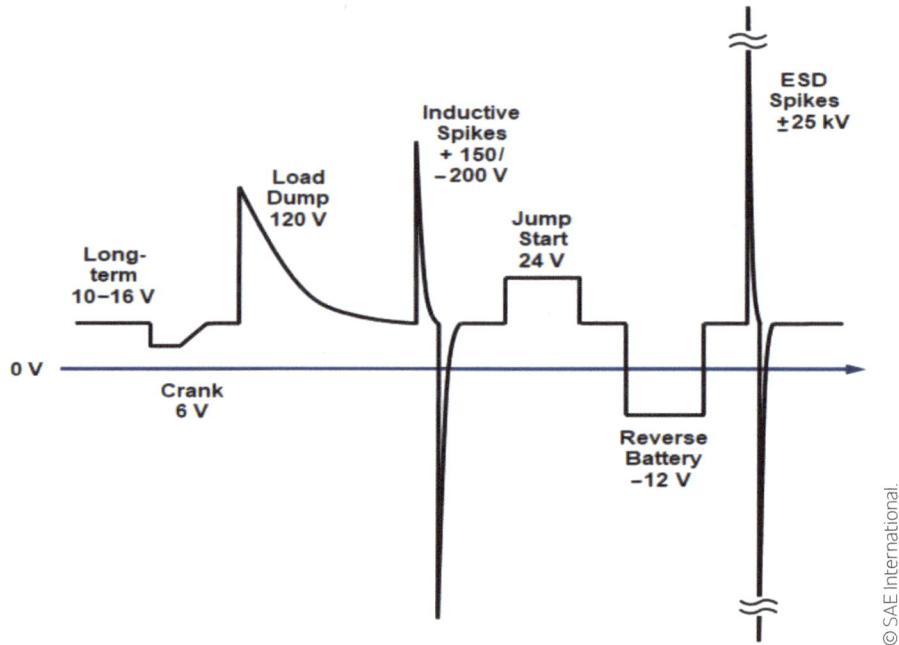

damage to electric loads, modules containing microcontrollers may experience software lockups, incorrect shutdowns, or incorrect restarts, potentially leading to unsafe operation of connected devices. For instance, a door lock motor left permanently on due to a software malfunction may overheat.

Load dump is a well-known phenomenon when the battery unexpectedly disconnects from the alternator while the engine is running. This can cause the system voltage to surge to approximately 100 V for several hundred milliseconds.

Inductive loads, abundant in cars, introduce positive and negative voltage spikes during connection and disconnection. These spikes can reach up to +150 V/−200 V and last approximately a hundred microseconds.

Jumpstart scenarios can impose double the voltage (i.e., 24 V) on the system for several

seconds, while reversing the battery connection can result in a −12 V supply. If electronic loads are not designed to withstand these conditions or protection fuses do not open, more extensive failures and internal short circuits may occur.

ESD is another significant transient that can damage car electronics, as it can reach up to ±25 kV for a few microseconds.

Lastly, strong electromagnetic fields, known as electromagnetic interference (EMI), can disrupt the regular operation of electronic modules or even cause damage. These interferences can enter the system through coupling into the 12 V supply and signal lines or as radiated interference affecting the electronic boards.

Every automotive manufacturer has specifications regarding these events, and international regulations are in place. The key takeaway is that failure to design electronic components to withstand these transients with sufficient margins can result in software and functional anomalies, component damage, and, in extreme cases, fire hazards.

4.8.
Fuse × Wire Gage

Electric fuses have served as valuable safeguards for several decades, mitigating the detrimental effects of overloads and short circuits [4.15]. **Figure 4.15** captures some examples. As discussed in numerous contexts, excessive current can lead to overheating and damage to wiring harnesses, connectors, and electronic components. The role of a fuse is to interrupt the electric current flow when it exceeds safe levels, thereby preventing further harm.

However, fuses are not capable of solving all electrical issues. There are specific scenarios and applications where fuses may prove ineffective or impractical. Also, proper selection of the wire gauge connected to each fuse must be made to ensure optimal protection. Otherwise, the wire may overheat in overload and catch fire while the fuse remains intact.

Fuses are typically not used in some specific applications, like the start motor feed, due to the substantial inrush current, which reaches several hundred amperes. Introducing a fuse in such a circuit would lead to a significant voltage drop, detrimental to the system performance.

4.9.
Short Circuits

Limited and dead short circuits, previously addressed regarding harness and connector degradation, can also manifest within electric modules and components. **Figure 4.16** depicts an illustrative instance of such a short circuit, where the damage is confined solely to an integrated circuit. Both forms of short circuits, when they occur in an electric part, can lead to significant harm and give rise to vehicle fires.

Figure 4.16 Burnt chip in a PCB.

Kapustin Igor/Shutterstock.com.

Figure 4.15 Some automotive fuses.

J.J. Gouin/Shutterstock.com.

4.10.
High-Voltage Arcing

Components and harnesses exposed to higher voltage, such as spark plug connectors, xenon lamp igniters, and xEV traction motors, require robust designs to mitigate the risk of arcing paths. **Figure 4.17** exhibits a burnt spark plug coil, in which the damage did not spread to other parts. Arcing can generally lead to carbonization of adjacent insulators, potentially resulting in

Figure 4.17 Burnt spark plug coil.

yunus_oz/Shutterstock.com.

elevated temperatures of these materials and posing a significant risk of vehicle fires.

4.11.
Lightning

Although statistically rare, lightning discharges can severely threaten automotive electric systems. When a discharge occurs, it results in a surge of high voltage and current, often causing significant damage to multiple modules and potentially leading to a fire outbreak. **Figure 4.18** illustrates a real-life example of a pickup truck that experienced such an incident on a highway, with the lightning entering through the antenna pole.

The electrical system endured substantial damage, resulting in a dashboard fire outbreak. Fortunately, the protective Faraday cage effect played a crucial role, enabling the driver and passenger to survive unharmed and swiftly exit the vehicle.

Figure 4.18 Pickup hit by lightning.

Courtsey of Mike Woodman - Richer Fire Department.

4.12.
Improper Accessory Installation

The proliferation of electric accessories available in aftermarket shops, ranging from high-intensity lamps to multimedia systems, has introduced significant risks to vehicle safety. Unfortunately, many of these accessories lack the necessary robustness to withstand typical vehicles' stresses, leading to eventual failure. Adding to the problem, some technicians and installers do not exercise due care during installation. They may resort to arbitrary cuts, use wire gauges insufficient to support regular fuse operation, improperly apply insulation tape, and neglect other crucial safety practices.

Over time, these substandard installations can result in functional failures and, in severe cases, pose a serious fire hazard. A real-life illustration of such recklessness can be seen in **Figure 4.19**, where an alarm was carelessly installed, putting the entire electric system at risk.

Figure 4.19 Careless installation of an alarm module.

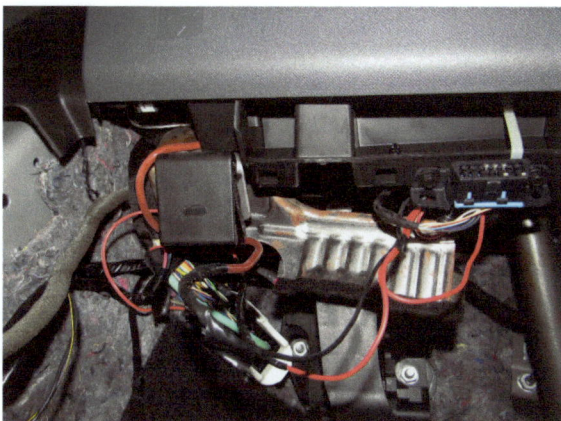

Courtesy of Erbis Llobet Biscarri.

References

[4.1].   Barnett, G., *Vehicle Battery Fires: Why They Happen and How They Happen*, SAE International Book R-443 (Warrendale: SAE International, 2017), ISBN:978-0-7680-8143-5.

[4.2].   Modi, M. and Galgoci, B., "Evaluation of High Resistance Connection in Automotive Application," *SAE Int. J. Adv. & Curr. Prac. in Mobility* 2, no. 5 (2020): 2751-2759, doi:https://doi.org/10.4271/2020-01-0926.

[4.3].   SAE International Ground Vehicle Standard, "Low Voltage Primary Cable," SAE Standard J1128, December 2020.

[4.4].   SAE International Ground Vehicle Standard, "Automobile and Motor Coach Wiring," SAE Standard J1292, April 2016.

[4.5].   DeMarois, P.H., Ballard, W., Engle, J., West, G. et al., "Full Scale Burn Demonstration of Two 2013 Ford Fusions – Arc Mapping Analysis," SAE Technical Paper 2018-01-1439 (2018), doi:https://doi.org/10.4271/2018-01-1439.

[4.6].   Mroczkowski, R., *Electrical Connector Handbook: Theory and Applications* (New York: McGraw Hill, 1998), ISBN:0-07-041401-7.

[4.7].   Kakuta, N., "Crimp Analysis Simulation Technology," *SAE Int. J. Passeng. Cars - Electron. Electr. Syst.* 1, no. 1 (2009): 501-507, doi:https://doi.org/10.4271/2008-01-1271.

[4.8].   Revie, R.W. and Uhlig, H.H., *Corrosion and Corrosion Control: An Introduction to Corrosion Science and Engineering*, 4th ed. (Hoboken: John Wiley & Sons, 2008), ISBN:978-0-471-73279-2.

[4.9].   Song, J., Wang, L., Zibart, A., and Kochet, C., "Corrosion Protection of Electrically Conductive Surfaces," *Metals* 2, no. 4 (2012): 450-477, doi:https://doi.org/10.3390/met2040450.

[4.10]. Park, Y., Jung, J., and Lee, K., "Overview of Fretting Corrosion in Electrical Connectors," *International Journal of Automotive Technology* 7, no. 1 (2006): 75-82.

[4.11]. Aruna Devi, D., Veeramanikandan, M., Vavilapalli, K.R., and Jeevan, N.K., "HiL Testing of Automotive Transients in Electric Vehicle," SAE Technical Paper 2022-28-0378 (2022), doi:https://doi.org/10.4271/2022-28-0378.

[4.12]. International Organization for Standardization, "Road Vehicles—Electrical Disturbances from Conduction and Coupling," Standard ISO 7637, 2011-2016.

[4.13]. International Organization for Standardization, "Road Vehicles—Test Methods for Electrical Disturbances from Electrostatic Discharge," Standard ISO 10605, 2023.

[4.14]. International Organization for Standardization, "Road Vehicles—Environmental Conditions and Testing for Electrical and Electronic Equipment," Standard ISO 16750, 2018.

[4.15]. SAE International Ground Vehicle Standard, "Blade Type Electric Fuses," SAE Standard J1284, April 1988.

Chapter 05

Other Degradation Mechanisms

5.1.
Introduction

Automotive components are exposed to many environmental factors and operational stresses, gradually undermining their structural integrity and functionality and eventually leading to fire hazards. Understanding the mechanisms contributing to the degradation of automotive parts is crucial for designing resilient and secure vehicles and is a necessary background for analyzing incidents. This chapter delves into various degradation mechanisms, ranging from thermal damages and exposure to infrared and ultraviolet radiation to the detrimental effects of water in its various forms and manifestations, aggressive chemicals, solid particles, accessories, aging, and ever-elusive random failures.

Exploring these factors aims to equip automotive engineers, researchers, and safety professionals with the knowledge to identify, prevent, and mitigate potential hazards, ensuring modern vehicles' continued safety and reliability.

5.2.
Ultraviolet (UV) Radiation

UV radiation emitted by the Sun has detrimental effects on various automotive parts and components over time, necessitating protective measures. The high-energy photons in UV radiation interact with the molecular structure of materials, particularly plastics and rubber, resulting in chemical and physical changes that compromise their performance and longevity.

5.2.1.
Plastics

Exposure to UV radiation initiates photooxidation, where UV rays and oxygen interact with and degrade the polymer molecules. The UV rays break down the molecular chains and generate free radicals. These radicals react with atmospheric oxygen, further modifying the chemical and physical properties of the material. Consequently, the plastic becomes discolored, brittle, and prone to cracking [5.1, 5.2].

5.2.2.
Rubber

UV radiation accelerates the aging process of rubber components, including seals, gaskets, and tires. The UV rays break down the chemical bonds within the rubber, causing a loss of elasticity and increased susceptibility to cracking and degradation. This deterioration can lead to leaks, reduced performance, and a shortened lifespan for these components.

These degradations of automotive parts impair their mechanical robustness, electrical properties, and sealing capabilities. These compromised functions can ultimately escalate into fire-related incidents. While UV stabilizers and additives are typically incorporated into materials during the production process to retard these effects, their occurrence cannot be eliminated.

5.3.
Infrared Radiation (IR)

IR plays a significant role in thermal transfer mechanisms and warrants specific attention due to its diverse effects. One notable situation is the greenhouse effect within a vehicle cabin when exposed to sunlight, which can lead to temperatures reaching 185°F (85°C) or higher. Moreover, IR is also responsible for heating and potentially degrading various components, connectors, and harnesses located in the irradiation field of exhaust collectors, catalytic converters, and turbochargers. The surface temperatures measured on these components can reach 1067°F (575°C), at which IR emission is significant, besides the risk of direct contact with flammable substances.

Therefore, cabin components and electronic modules are designed to withstand the required temperatures. Thermal shields are strategically employed between the radiation source and adjacent components to mitigate the adverse impacts of IR generated by hot parts. These shields effectively reduce the temperature of nearby parts, ensuring their longevity and optimal performance.

5.4.
Sunlight

Although direct sunlight exposure alone typically cannot raise the temperature of combustible materials inside vehicles to the point of ignition, incidents involving the unintended use of lenses have been documented, altering this situation. Placing a plastic bottle filled with water or transparent liquid on a car seat can lead to this scenario. On a sunny summer day, when the illumination angle is just right and the focal point of the unexpected lens aligns with the seat surface, an easily flammable seat fabric could result in a dangerous combination with potentially catastrophic consequences [5.3].

5.5.
Thermal Damage

Besides IR exposure, other thermal-related events can significantly impact automotive components, leading to damage or degradation over time. These events can manifest in various ways, such as extreme temperatures leading to brittleness or melting, plastic deformation of polymers, thermal expansion/contraction stresses, and thermal cycles leading to fatigue [5.4].

5.5.1.
Extremely Cold Temperatures

At very low temperatures, certain materials become brittle and prone to cracking. The low temperatures can cause a reduction in ductility and toughness, making the components more susceptible to fracture or failure under stress or impacts. The brittle behavior can lead to sudden and unexpected failures. For example, in cold climates, rubber hoses can become stiff, losing flexibility and increasing the risk of rupturing or snapping.

5.5.2.
Extremely Hot Temperatures

When exposed to high temperatures, plastic and rubber parts can experience softening, melting, or degradation. The components can deform, lose their shape, and potentially fail. Melting may start as low as 167°F (75°C) for specific blends of polyvinyl chloride (PVC), for instance. Moreover, the insulation polymer used in electrical wiring and connectors may also suffer damage or insulation breakdown when exposed to excessive heat, resulting in electrical failures or short circuits.

Also, high temperatures often accelerate metal oxidation, which can degrade electric

connections and contacts. These situations can occur in engine compartments, next to exhaust systems, or in other areas exposed to intense heat [5.5, 5.6].

5.5.3.
Cooling System Failures

Automotive engines, transmissions, and other systems rely heavily on cooling mechanisms to regulate their operating temperature inside a safe area. When the cooling system fails, it can result in severe consequences such as engine overheating. This, in turn, can cause detrimental effects like warping cylinder heads or blowing gaskets, posing a potential fire hazard if flammable fluids contact hot engine components. Additionally, failure in the cooling system can lead to lubrication degradation, resulting in increased friction, overheating, reduced lifespan, and ultimately catastrophic failures.

Furthermore, cooling systems are crucial for other modern powertrain alternatives, such as traction batteries, high-power electric motors, and electronics. If their temperature limits are exceeded, these components are also susceptible to damage and safety hazards.

The effectiveness of the cooling system can be compromised if its fluid leaks through faulty hoses, radiators, reservoirs, or connections. Excessive pressure is another potential cause of cooling system failure, often caused by a malfunctioning radiator cap or a faulty pressure relief valve. Such extreme pressure can lead to coolant leaks through damage to its hydraulic circuit.

5.5.4.
Plastic Deformation of Polymers

Many parts contain polymers, such as seals, gaskets, and connectors. When subjected to elevated temperatures—yet below the melting point—these polymers can undergo plastic deformation, which results in a permanent change in shape and mechanical properties. Over time, this can lead to a loss of sealing capabilities or reduced functionality of the affected components.

5.5.5.
Expansion/Contraction Stresses

Automotive components are regularly subjected to temperature changes during operation. When exposed to heat, materials tend to expand and will contract when cooled. This thermal expansion and contraction can create stresses within the components. When dissimilar materials are attached, the difference in their thermal expansion coefficients will induce stress. Also, when large parts are exposed to thermal shocks, the expansion and contraction gradients generate internal tensions. If these stresses exceed the material's capacity to withstand them, they can lead to distortion, warping, delamination, or even cracking of the parts.

Moreover, large electronic modules that are sealed will face several cycles of internal pressure rising and falling as their interior air expands and contracts. If not properly vented, this might damage their housing or sealing and allow water intrusion. Chapter 6 will examine this mechanism in more detail.

5.5.6.
Thermal Cycles Leading to Fatigue

Automotive components are repeatedly subjected to thermal cycles since they are manufactured. These cycles involve heating and cooling due to engine operation, braking, and ambient temperature changes. Over time, these thermal cycles can lead to fatigue failure. Fatigue occurs when repeated heating and cooling cause

microstructural changes in the materials, leading to the initiation and propagation of cracks. Eventually, these cracks can grow and result in component failure, particularly in areas of high-stress concentration or with inherent material defects.

5.6.
Exhaust System Failures

Modern vehicle exhaust systems have evolved into intricate and sophisticated structures, introducing new challenges regarding potential failures that may lead to car fires. These systems' complex design and functionality, featuring numerous hot components capable of igniting combustible substances upon contact, coupled with the eventual release of highly heated gases due to wear and tear, significantly heighten the risk of fire incidents.

Gasoline engines commonly incorporate one or more catalytic converters to reduce emissions. Despite their effectiveness, these converters operate at elevated temperatures, increasing the potential for igniting flammable materials upon direct contact or proximity to the hot exhaust gases. EGR valves, utilized in various engines to enhance fuel efficiency and reduce emissions, pose similar fire hazards if unforeseen malfunctions allow combustible substances to make contact with or approach the EGR or if EGR gases escape.

In the case of contemporary diesel engines, exhaust systems have become more complex. Periodically, some diesel engines burn accumulated soot by injecting unburnt diesel into a specialized exhaust system filter near the engine.

This filter experiences repeated temperature surges during the soot cleaning cycle, close to 1100°F (approximately 600°C). Many diesel systems also integrate a selective catalytic reduction (SCR) system, applying a fluid comprising water and urea (AdBlue™ or Arla 32™) to a hot ceramic filter. This process converts NOx emissions and other pollutants into nitrogen, water vapor, and carbon dioxide, with the ceramic filter operating at approximately 650°F (about 350°C).

Both gasoline and diesel engines often enhance their efficiency with turbochargers, utilizing engine exhaust gases to drive a high-speed turbine connected to a cold turbine, increasing the air pressure injected into the cylinders. The presence of a wastegate, located either within the turbocharger or adjacent to it, serves to redirect excess pressure away from the hot turbine, optimizing engine boost control. However, these turbocharger components and other hot parts introduce similar fire risks in the event of failures or when combustible materials come into proximity.

5.7.
Water

Water is a critical substance that needs careful consideration due to its presence in various forms and environments. From humidity to heavy rain and even seawater, water on automotive components and systems can lead to functional failures and damage if not adequately addressed. Therefore, designing, manufacturing, and servicing vehicles to withstand these conditions

is paramount. In complement to the concepts examined here, Chapter 6 will focus more on the relationship between water and electronics.

5.7.1.
Humidity

Even in areas with relatively low rainfall, humidity can still pose a challenge for automotive components. High humidity levels can lead to corrosion and oxidation of metal parts and connections, compromising the performance and longevity of critical systems.

5.7.2.
Condensation

Condensation will occur when warm and moist air contacts colder surfaces within the vehicle. This can lead to the formation of water droplets on windows and internal components, such as the metallic cross-car beam under the instrument panel. In some cases, excessive condensation may result in these droplets precipitating over electronic components and their subsequent failure.

5.7.3.
Drizzle and Rain

From drizzles to severe rain, water can infiltrate exposed areas of a vehicle, potentially seeping into electrical connections, engine compartments, and other sensitive regions. Also, while driving at high speed on wet pavements, the cloud created by other vehicles, as captured in **Figure 5.1**, will be ingested by the engine ventilation system and reach many underhood parts. These infiltrations, if not proactively anticipated and safeguarded against, can lead to malfunctions, damage to electronic components, and short circuits.

Figure 5.1 Truck creating a water cloud.

LanaElcova/Shutterstock.com.

5.7.4.
Snow and Ice

Snow poses a unique challenge for automotive systems in colder climates. As seen in **Figure 5.2**, accumulated snow and ice can obstruct various vehicle parts, including actuators, vents, and drains. Moreover, the subsequent melting can lead to water ingress, increasing the risk of electrical failures.

Figure 5.2 Engine covered with snow.

Aleksey R/Shutterstock.com.

5.7.5.
Washing

While washing vehicles is routine, it can pose a risk if water enters sensitive components such as electronic modules or electrical connections. Improper washing techniques or high-pressure water jets in sensitive areas can cause damage. Washing the engine compartment with high-pressure water jets, as captured in **Figure 5.3**, can lead to water intrusion to sensitive electronic connectors and subsequent damage.

Figure 5.3 Do not do that with your car.

boyphare/Shutterstock.com.

5.7.6.
Pools in Potholes and Flooding

Vehicles driving through water-filled potholes and flooded areas will experience splashing or water entry into the undercarriage, potentially damaging the vehicle's electrical and mechanical systems, like other water-related damage. **Figure 5.4** shows an example of a car splashing muddy water on a flooded road.

Figure 5.4 Car on a flooded road.

Kravtzov/Shutterstock.com.

5.7.7.
Sea Water

Vehicles near coastal regions, those exposed to seawater, or those impacted by flooding require special attention. Seawater is highly corrosive and can rapidly degrade metal surfaces, electrical connections, and seals.

Components and systems must be designed, manufactured, and serviced to withstand all these water-related conditions, as they can lead to functional failures and damage. Chapter 6 will examine the importance of water in electric systems and watertightness in more detail.

5.8.
Aggressive Chemicals

Aggressive chemicals pose a significant risk to the integrity and performance of vehicles' hoses, harnesses, and electrical systems. The interaction between these chemicals and automotive components can lead to failures and potentially escalate to fire hazards.

5.8.1.
Engine Coolant

One widespread aggressive chemical found in vehicles is engine coolant, which typically contains ethylene glycol or propylene glycol. While essential for regulating engine temperature, these chemicals can have detrimental effects when they encounter hoses and electrical components. Over time, coolant can cause degradation of the rubber compounds used in some hoses and sealings, making them brittle, prone to cracking, or causing leaks.

5.8.2.
Brake Fluid

Brake fluid, typically glycol ether-based, is another aggressive chemical that can corrode rubber hoses and seals upon contact. Also, if brake fluid leaks onto electrical components, it can damage wire harnesses and connectors, leading to potential malfunctions or short circuits.

5.8.3.
Fuel

Fuels, like gasoline and diesel, also threaten automotive components due to the presence of additives and compounds that degrade rubber hoses. This degradation can lead to fuel leaks, which are highly flammable and pose a direct fire hazard.

5.8.4.
Acids

Lead–acid battery electrolyte is another substance of concern due to its sulfuric acid content. This electrolyte can corrode metals and other materials, posing a risk to various automotive components. Figure 5.5 exhibits an example where one of the battery poles and its connection are visibly corroded. Also, some low-quality cleaning substances contain acids in their formulation and should not be used to clean a car.

Figure 5.5 Acid corrosion on battery pole.

sharkmonkey/Shutterstock.com.

5.8.5.
Salt and De-Icing Chemicals

In specific environments such as seashore locations or winter conditions where sea salt, road salt, and de-icing chemicals are present, corrosion becomes a significant issue for exposed metal parts and electrical connectors.

5.8.6.
Other Substances

Other aggressive chemicals in the automotive environment include oils, solvents, and cleaning agents. These substances can cause rubber hoses and seals to deteriorate, leading to leaks or loss of functionality. Sometimes, they can damage the insulation on electrical wires and connectors, potentially causing short circuits or electrical malfunctions.

5.9.
Solid Particles

The exposure of automotive parts to excessive dust, seeds, and other particles can harm the vehicle performance, including the degradation of lubrication and fire hazards. The frequency of exposure is particularly relevant in agricultural tractors, machinery, and off-road vehicles. The issues arise due to the reasons discussed in the following sections.

5.9.1.
Compromised Lubrication

Vehicles rely heavily on well-lubricated moving parts to ensure smooth operation and longevity. Lubricating oils and greases create a protective film that minimizes friction and prevents metal-to-metal contact. However, when dust and particles infiltrate the lubrication system, they disrupt the formation and maintenance of this essential protective film. Consequently, the moving components experience increased friction, heat, and wear, which can also lead to catastrophic failures. If neglected for an extended period, the ripped boot, as shown in **Figure 5.6**, will allow the entrance of dust and grit in this constant velocity (CV) joint, impair its lubrication, and shorten the joint life or even allow disastrous damage.

Figure 5.6 Ripped CV boot.

Courtesy of Erbis Llobet Biscarri.

5.9.2.
Combustible Foreign Materials

Seeds, leaves, and other organic materials in the engine compartment or close to the exhaust system pose a direct fire hazard, especially in agricultural tractors and machinery [5.7]. Also, in vehicles unattended for several days, rodents might nest in the engine compartment, bringing pits and other flammable substances, as illustrated in **Figure 5.7**. If these combustible materials meet hot engine surfaces or the exhaust system, they have the potential to ignite and trigger a fire.

Figure 5.7 Rat nesting in the engine compartment.

Kamonkanok/Shutterstock.com.

Careless maintenance operations can also lead to hazardous situations when combustible materials are left near hot components. Let us consider the scenario of an oil filter change where a piece of a wiping cloth is inadvertently left trapped in the lower region of the engine. As the vehicle is used afterward, the airflow generated may displace this cloth, causing it to meet the hot catalytic converter, thereby initiating a fire.

5.10.
Improper Installation of Accessories and Modifications

In Chapter 4, the risks associated with vehicle electric accessories have been already discussed.

However, knowing other accessories and modifications that can compromise a vehicle's safety and lead to potential fire hazards is important.

Some modifications, commonly known as tuning, involve altering the software of the engine control module to boost power and torque. Unfortunately, these enhancements often overlook their impact on emissions, engine temperatures, stresses, and overall service life.

Another example of such modifications is the engine's conversion to LPG, as exemplified in **Figure 5.8**. This conversion adds alternative fuel storage, delivery, and control systems, introducing inherent risks and safety concerns.

Figure 5.8 Close up of an engine converted to LPG.

sima/Shutterstock.com.

Moreover, some vehicle owners may add an engine top cover to improve aesthetics. Still, this seemingly harmless addition can inadvertently obstruct ventilation and create dangerous hot spots within the engine compartment.

Another modification that can have harmful effects is the installation of ballistic armor. It involves reinforcing and replacing various parts of the vehicle's structure and windows with materials designed to resist ballistic threats. While enhancing the safety and security of the occupants, this modification may have unintended consequences on other components and systems.

During installation, wiring harnesses and electrical connectors may be misplaced, stressed, cut, or smashed, leading to potential damage, misconnections, or short circuits. Additionally, the placement of ballistic armor may require alterations to the vehicle's fuel system, such as rerouting fuel lines or modifying the fuel tank. Care in executing these modifications is crucial

to avoid potential leaks or damage that could pose a safety hazard.

From a fire risk perspective, thoroughly evaluating, implementing, and regularly servicing these modifications and additions are of utmost importance. Negligence in this regard could lead to disastrous events and potential fire-related incidents.

5.11.
Inappropriate Service

It is crucial to ensure that vehicle servicing is conducted by trained professionals who follow manufacturer guidelines and industry standards and use the correct tools. Regular maintenance and servicing should be performed to uncover and address potential safety issues, ensure the proper functioning of the vehicle, and mitigate the risk of fire or other hazards.

Faulty electrical work during vehicle service, such as incorrect wiring or improper installation of electrical components, can lead to electrical system malfunctions. For example, if wires are not properly insulated or secured, they can rub against each other or other vehicle parts, resulting in exposed wires and potential sparks.

Mishandling or improper installation of fuel system components, such as fuel lines, filters, or injectors, can lead to fuel leaks. For instance, if a fuel line is not correctly connected or is damaged during service, it may result in fuel spraying or dripping onto hot engine components or electrical parts, leading to a fire.

During vehicle service, if seals, gaskets, or drain plugs are not correctly installed or tightened, it can result in oil or fluid leaks. Oil leaks can lead to the accumulation of flammable substances in the engine compartment or other parts of the vehicle, increasing the risk of fire.

If engine maintenance, including oil changes, cooling system verification, filter replacements, or belt adjustments, is not performed correctly, it can lead to engine malfunctions. This can result in decreased performance, overheating, or even complete engine failure.

It is important to note that these are just a few examples, and other safety and fire hazards arise from improperly conducted vehicle service. Regularly scheduled maintenance and adhering to the manufacturer's recommendations help ensure the safety and reliability of a vehicle. Also, most electronic modules come with self-diagnosis, which, if adequately accessed, can efficiently identify ignored issues and prevent more significant failures.

5.12.
Aging and Random Failures

Numerous factors that can diminish the expected lifespan of automotive parts have been examined, such as exposure to extreme temperatures, corrosive atmospheres, and mechanical overloads. Even without these factors, aging is inevitable, and random failures can occur during normal usage. Figure 5.9 illustrates an example where a hose's external rubber layer is ruined, necessitating prompt replacement. Considering these and other factors, adhering to regular service intervals and performing preventive maintenance correctly are essential. These practices help minimize the risk of safety-related incidents, even if the vehicle driver does not notice any decline in performance.

Figure 5.9 Damaged hose.

Shamils/Shutterstock.com.

References

[5.1]. Celina, M. and Assink, R., *Polymer Durability and Radiation Effects*, ACS Symposium Series (Washington, DC: American Chemical Society, 2007), ISBN:978-0-8412-6952-1.

[5.2]. Margolis, J., *Engineering Plastics Handbook*, 1st ed. (New York: McGraw-Hill Professional, 2005), ISBN:978-0071457675.

[5.3]. CBS News, "Warning: Leaving Bottled Water in Your Car Could Start a Fire," accessed November 2023, https://www.cbsnews.com/news/that-bottled-water-in-your-car-could-start-a-fire-firefighters-warn/.

[5.4]. Suresh, S., *Fatigue of Materials*, 2nd ed. (Cambridge, UK: Cambridge University Press, 2012), doi:https://doi.org/10.1017/CBO9780511806575.

[5.5]. Colwell, J., "Ignition of Combustible Materials by Motor Vehicle Exhaust Systems - A Critical Review," *SAE Int. J. Passeng. Cars - Mech. Syst.* 3, no. 1 (2010): 263-281, doi:https://doi.org/10.4271/2010-01-0130.

[5.6]. Colwell, J. and Biswas, K., "Steady-State and Transient Motor Vehicle Exhaust System Temperatures," *SAE Int. J. Passeng. Cars - Mech. Syst.* 2, no. 1 (2009): 206-218, doi:https://doi.org/10.4271/2009-01-0013.

[5.7]. Morse, T., Cundy, M., and Kytomaa, H., "Vehicle Fires Resulting from Hot Surface Ignition of Grass and Leaves," SAE Technical Paper 2017-01-1354 (2017), doi:https://doi.org/10.4271/2017-01-1354.

Water × Electricity

6.1.
Introduction

As seen in previous chapters, the presence of water must be considered in all its physical states and natural manifestations to ensure the intended vehicle's performance and safety over time.

In the dynamic realm of automotive technology, safeguarding against water-induced damage is paramount. This chapter offers a more detailed exploration of the challenges arising from the presence of water in the electric systems, encompassing electrolysis, dendritic growth, and other critical concerns. Understanding the electrochemical reactions, intrusion mechanisms, and damage from the interaction of water with electrical fields or currents is essential to ensure vehicles' sustained performance and safety.

This chapter also delves into the significance of Ingress Protection classification (IPXX) and sealing degradation mechanisms, which determine a vehicle's resistance to water and other contaminants and address factors like capillarity and pumping/respiration contributing to water infiltration and potential damage. By unraveling the complexities of these processes, it aims to equip the readers with valuable knowledge, enabling them to develop effective strategies to fortify automotive systems against the ever-present threats posed by water. Moreover, understanding these potential root causes becomes instrumental in incident analysis.

6.2.
Electrolysis/Electrolytic Corrosion/Dendritic Growth

Electrolysis, electrolytic corrosion, and dendritic growth are all interrelated electrochemical processes, but they have distinct characteristics and occur under different circumstances. However, when these mechanisms arise, a certain level of degradation will eventually evolve into failures or fire hazards. Let us examine each process.

6.2.1.
Electrolysis

Electrolysis refers to using an electric current to drive a nonspontaneous chemical reaction. It involves decomposing an electrolyte, a conducting medium like salts dissolved in water, by applying an electric current. This process occurs in an electrolytic cell, as depicted in **Figure 6.1**: two electrodes, an anode (positive) and a cathode (negative), are immersed in the electrolytic solution. The anode attracts negative ions (anions) and undergoes oxidation, while the cathode attracts positive ions (cations) and undergoes reduction. As electrons flow through the external electric circuit, ions flow within the electrolytic solution.

This process often removes metal ions from the anode, causing corrosion. The ions then transfer through the electrolyte to the cathode. In some cases, significant amounts of gases, such as oxygen and hydrogen, may be generated, mainly when the electrolytic solution contains water and current densities are high. In the automotive environment, undesired electrolytic cells may occur when energized wires or conductors are exposed to water with contaminants or other automotive fluids.

Figure 6.1 Electrolysis experiment.

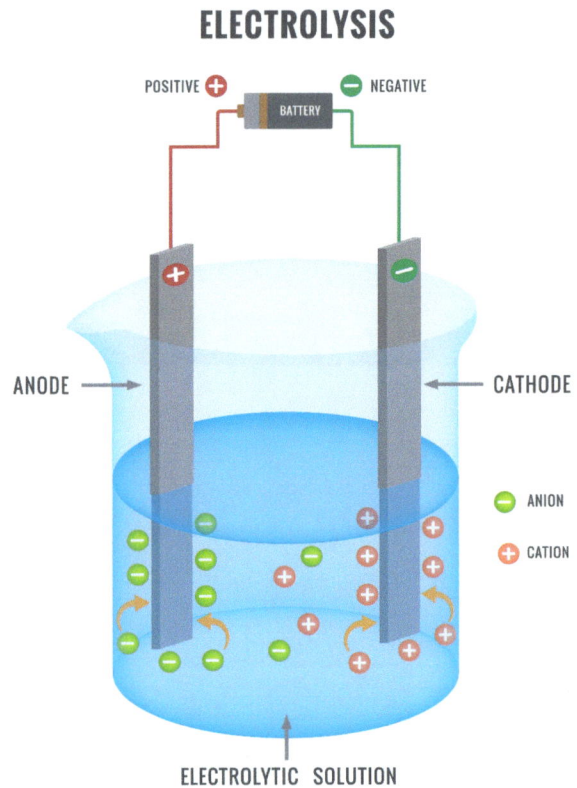

6.2.2.
Electrolytic Corrosion

Electrolytic, galvanic, or bimetallic corrosion occurs when two dissimilar metals or alloys are in contact with each other and in the presence of an electrolyte, such as water or a humid environment. The difference in electrochemical potentials between the two metals leads to the spontaneous flow of electric current, resulting in accelerated corrosion of the less noble (more reactive) metal. This type of corrosion can cause damage to the metal surfaces [6.1, 6.2]. It can be problematic, especially in connections using different metals or metal plating, as exemplified by the battery pole connection shown in **Figure 6.2**.

Figure 6.2 Electrolytic corrosion on a battery pole.

Taphat Wangsereekul/Shutterstock.com

6.2.3.
Dendritic Growth

Dendritic growth is forming tiny branching, tree-like structures during electrochemical deposition processes. It occurs when metal ions or solute particles migrate between two closely spaced electrodes under the influence of an electric field and is also known as electro-chemical migration. Moisture and condensation may be sufficient to form the electrolyte, even without a visible liquid bridge between the electrodes. As ions accumulate over one electrode, they form dendritic structures that extend into the surrounding medium, reaching the other electrode. **Figure 6.3** magnifies an example of dendrites growing between two printed circuit board tracks a fraction of an inch (a few millimeters) apart. Sometimes, dendritic growth can cause leakage currents that intensify over time, leading to electrical malfunctions and short circuits [6.3].

Figure 6.3 Dendrites between electronic board tracks.

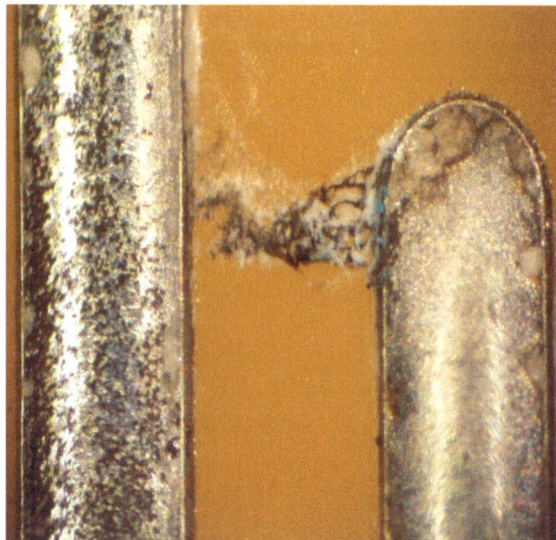

Xiaofei He, Michael H. Azarian, Michael G. Pecht, Analysis of the Kinetics of Electrochemical Migration on Printed Circuit Boards Using Nernst-Planck Transport Equation, Electrochimica Acta, Volume 142, 2014, Pages 1-10, ISSN 0013-4686, Permission granted by Elsevier through Copyright Clearance Center.

Note: Chapter 8 will discuss the dendritic growth inside lithium batteries in more detail.

6.3.
IPXX

The IPXX is a standardized measure that defines the degree of protection offered by an equipment enclosure against solids and liquids. Initially established by the International Electrotechnical Commission (IEC) for electric equipment, even some mechanical components have adopted these criteria. The abbreviation "IP" stands for Ingress Protection, and the first "X" digit, when replaced by a number, indicates the level of protection against solids, as outlined here [6.4, 6.5]:

0: No protection against contact and ingress of objects.

1: Protection against solid objects larger than 50 mm, such as a hand.

2: Protection against solid objects larger than 12.5 mm, such as fingers.

3: Protection against solid objects larger than 2.5 mm, such as tools and wires.

4: Protection against solid objects larger than 1mm, such as small tools and wires.

5: Dust-protected. Dust ingress is not entirely prevented but does not interfere with the equipment's satisfactory operation.

6: Dust-tight. No ingress of dust is permitted, providing complete protection.

Similarly, the second X digit indicates the level of protection against liquids:

0: No protection against liquids.

1: Protection against vertically falling drops of water or condensation.

2: Protection against vertically falling drops of water when the enclosure is tilted up to 15°.

3: Protection against water sprays at angles up to 60° from vertical.

4: Protection against water splashes from all directions.

5: Protection against low-pressure jets of water from all directions.

6: Protection against powerful jets of water or heavy seas.

7: Protection against the effects of temporary immersion in water up to 1 m for 30 min.

8: Protection against continuous immersion in water under conditions specified by the manufacturer.

9: Protection against high-pressure jets of water and steam cleaning.

For instance, an enclosure with an IP65 rating signifies that it is dust-tight (level 6) and protected against low-pressure water jets from any direction (level 5). In the automotive industry, numerous components are specified according to an IPXX index to ensure the necessary protection.

6.4.
Sealing Degradation Mechanisms

While the correct specification of an IP category initially provides the necessary protection for automotive modules, this shield level might degrade over time due to several factors. The following sections discuss some potential reasons for the degradation.

6.4.1.
Vibration and Mechanical Stress

Frequent vibrations and mechanical stress experienced during vehicle operation can loosen or damage protective seals, locks, and housings. This can create gaps or openings that allow the entry of dust, water, or other foreign particles, compromising the component's IP rating.

6.4.2.
Temperature Variations

Exposure to a wide range of temperatures, including extreme heat and cold, thermal shocks, and repeated cycles, causes materials to expand and contract, leading to volumetric changes that, in turn, generate mechanical stresses. These stresses can degrade seals or joints over time, potentially compromising the component's ability to maintain its IP classification.

6.4.3.
Exposure to Chemicals and Corrosion

Automotive components may encounter corrosive substances like road salt, chemicals, and atmospheric pollutants. Corrosion can weaken the enclosure structure, erode surfaces that mate with seals, or directly attack the sealing material, creating gaps or cracks through which contaminants can enter.

6.4.4.
Improper Maintenance

Inadequate or improper maintenance practices can also contribute to the degradation of the IP rating or expose parts to unplanned situations. For instance, using high-pressure water jets, underhood may force water into sensitive connectors and other components. Moreover, using sharp metallic needles to probe electric circuits without due caution poses a significant risk of harming connector seals or compromising the insulation layer of wires. Regrettably, some technicians deliberately pierce the wire insulator, aiming to establish contact with the copper wire within, employing sharp probes like the one depicted in **Figure 6.4**. Additionally, the failure to conduct regular inspections and the neglect to replace damaged or worn-out seals and covers can further expose components to the infiltration of contaminants.

6.4.5.
Accidental Damage

Accidental drops and impacts on components and vehicle collisions can cause physical damage to automotive parts, including their protective enclosures. Cracks or openings resulting from such incidents can significantly reduce the IP index.

6.4.6.
Wear, Tear, and Aging

The inevitable aging of a vehicle and repeated exposure to expected environmental conditions, such as extreme temperatures, UV radiation, IR radiation, vibrations, and mechanical stresses, can cause physical damage to the component's protective enclosure. Over time, this wear and tear can compromise the enclosure integrity and reduce its ability to prevent the ingress of solids and liquids. Rubber seals and gaskets can become brittle and lose their elasticity or soften and deform, compromising their ability to provide a tight seal. **Figure 6.5** exemplifies an underhood fuse and relay box that, despite originally having a protective cover, was damaged by water ingress.

Figure 6.4 Needle probe.

rCarner/Shutterstock.com.

Figure 6.5 Fuse box spoiled by water.

DROPERDER/Shutterstock.com.

6.5.
Capillarity

Capillarity is a fascinating phenomenon that enables liquids, like water, to defy gravity and ascend in narrow spaces or tubes, exemplified by the glass tubes in **Figure 6.6**. In this context, the thinner the tube, the higher the water meniscus. This movement results from the combined adhesive and cohesive forces between the liquid and the solid surface, with Jurin's law defining the height of the water meniscus. According to this law, the water height within a capillary tube made of glass or copper is inversely proportional to the tube's diameter [6.6].

Figure 6.6 Capillary tubes.

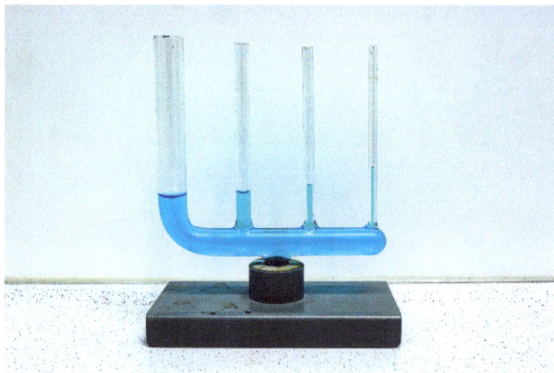

Menno van der Haven/Shutterstock.com.

To ensure flexibility, automotive wires consist of multiple copper strands twisted together, as shown in **Figure 6.7**. These strands have a circular cross section, leaving small spaces or capillaries between them, depicted as white spaces in the cross section shown in **Figure 6.8**, which are typically filled with air. These capillaries can serve as channels for water to travel through if it enters the wire from one of its ends or through a hole in its plastic insulator.

Figure 6.7 Exposed strands of an automotive wire.

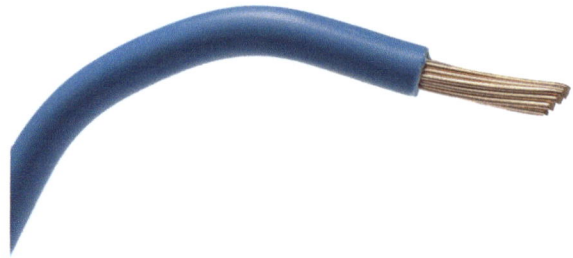

IB Photography/Shutterstock.com.

Figure 6.8 Automotive wire cross section.

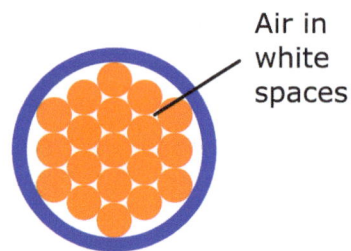

Air in white spaces

© SAE International.

As the water molecules adhere to the surfaces of the copper strands and form a continuous column, they can move upward against gravity due to cohesive forces. This upward movement is often referred to as capillary rise. The narrower the gaps between the strands, the greater the capillary elevation can be. In the case of automotive wires, the capillary action can allow water to travel several feet (various decimeters) upward, reaching distant connectors and electronic modules located higher than the point of entry of the water. This can occur because the capillaries within the wire strands extend throughout the length of the wire, enabling water to travel continuously along them.

The practical consequence is that connector harnesses cannot be left open in a vehicle if their extremity might be exposed to water. Consider the case of a fog light connector left open: as it is close to the ground when the vehicle travels on a wet floor, water might be splashed in this connector, move along the harness through the capillary action, and reach the electronic module that commands the fog light, damaging it.

6.6.
Pumping/Respiration

In addition to capillarity, water can be transported along a regular automotive wire through pumping or respiration [6.7]. Consider the setup illustrated in **Figure 6.9(a)**: a typical electronic module with an electronic board surrounded by air, a connector CN1, and a plastic case that provides at least IP64 protection, that is, it is dust-proof and safeguarded against water splashes in all directions. The module has a cable attached via a sealed connector, CN2, while the connector at the other end of the harness, CN3, is exposed to water.

When the electronic module heats, its internal air expands and flows through the interstitial spaces between the wire strands, bubbling to the atmosphere via CN3, as detailed in **Figure 6.9(b)**. Although the harness and its connectors prevent water ingress from the exterior, there is usually no sealing between the air inside the module and the

terminals of the board connector CN1 or between the terminals of the harness connectors CN2 and CN3 and the interior of each wire. Therefore, warm, expanded air inside the electronic module can easily flow through this unintended pneumatic circuit. Once thermal equilibrium is reached, the airflow naturally ceases.

However, when the module cools down, as shown in **Figure 6.9(c)**, the air inside contracts, reducing the air pressure along the wires and drawing water from connector CN3 into the wires until thermal equilibrium is achieved again. Due to the surface tension effect, this water will be entrapped inside the cables. Each warming/cooling cycle of the electronic module progressively moves the waterfront toward the module until it reaches CN1/CN2. Electrical failures or electrolytic corrosion will manifest at this point, eventually leading to catastrophic results.

In real-life scenarios, even if CN3 is not entirely immersed in water but is subjected to splashes or droplets during rain, water will gradually be pumped toward the electronic module. The capillary effect will further spread the water upward, accelerating the movement of the waterfront.

Figure 6.9 Pumping mechanism.

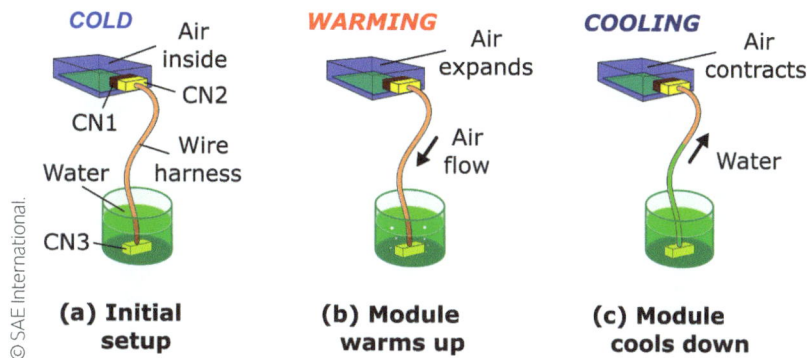

(a) Initial setup   **(b) Module warms up**   **(c) Module cools down**

To overcome this problem, a few solutions exist. One approach is avoiding the exposure of any unused connector to water. Another option is using sealed wires, where the interstitial spaces between the wire strands are filled with a sealing elastomer, despite the higher cost of these wires. Additionally, the application of a respiration valve, highlighted in **Figure 6.10**, proves effective [6.8, 6.9]. Its crucial element is a membrane made of polytetrafluoroethylene (PTFE) tissue with tiny pores, allowing for the free flow of air and continuously balancing air pressure inside the module. Simultaneously, this membrane prevents water droplets from crossing it, ensuring the maintenance of an adequate IPXX rating.

Figure 6.10 Module with ventilation valve.

Courtesy of Erbis Llobet Biscarri.

References

[6.1]. Vignesh, E. and Sundararajan, T., "Corrosion Prevention and Surface Engineering Practices in Auto-Components Industries," in: Kamachi Mudali, U., Subba Rao, T., Ningshen, S., Pillai, R.G. et al. (eds.), *A Treatise on Corrosion Science, Engineering and Technology*, Indian Institute of Metals Series (Singapore: Springer, 2022), https://doi.org/10.1007/978-981-16-9302-1_21.

[6.2]. Zhao, D., Liu, M., Li, G., Tan, C. et al., "Automobile Corrosion Mechanism and Cases Analysis," in: (SAE-China), S. (eds.), *Proceedings of the 19th Asia Pacific Automotive Engineering Conference & SAE-China Congress 2017: Selected Papers. SAE-China 2017*, Lecture Notes in Electrical Engineering, Vol. 486 (Singapore: Springer, 2019), https://doi.org/10.1007/978-981-10-8506-2_18.

[6.3]. He, X., Azarian, M., and Pecht, M., "Comparative Assessment of Electrochemical Migration on Printed Circuit Boards with Lead-Free and Tin-Lead Solders," *SMT Magazine*, Center for Advanced Life Cycle Engineering (CALCE), University of Maryland, August 2009.

[6.4]. International Electrotechnical Commission, "Degrees of Protection Provided by Enclosures (IP Code)," International Standard IEC 60529 (2.2 ed.), 2013, ISBN:9782832210864.

[6.5]. International Organization for Standardization, "Road Vehicles—Degrees of Protection (IP Code)—Protection of Electrical Equipment against Foreign Objects, Water, and Access," Standard ISO 20653:2013, 2013.

[6.6]. De Gennes, P.G., Brochard-Wyart, F., and Quéré, D., "Capillarity and Gravity," *Capillarity and Wetting Phenomena* (New York: Springer, 2004), 33-67, https://doi.org/10.1007/978-0-387-21656-0_2.

[6.7]. LEC, "La remontée d'eau par le cable d'alimentation" (Pumping Effect and Capillary Action in Cables), LEC-Lyon, accessed march 2024, https://www.lec-expert.com/topics/waterproof-led-lights--preventing-capillary-action-in-power-supply-cables.

[6.8]. Donaldson, "Donaldson Ventilation Solutions," accessed July 2023, https://www.donaldson.com/content/dam/donaldson/venting/literature/north-america/industries-markets/automotive/f118522-eng/Integrated-Venting-Solutions-Capabilities-for-Automotive-Applications.pdf.

[6.9]. Nitto, "Temish Ventilation Cap," accessed July 2023, https://www.nitto.com/us/en/products/temish_search/003/.

xEV Fundamentals

7.1.
Introduction

The automotive industry is currently undergoing a significant transformation driven by the introduction of electrified propulsion vehicles. This includes a spectrum of options ranging from BEVs to HEVs and FCEVs. This chapter explores the diverse configurations and acronyms that characterize this new era, organized alphabetically for easy reference [7.1-7.7].

It delves into the distinctions between various types of xEVs and the subcategories associated with HEVs, each embodying a unique blend of electric and traditional combustion propulsion. As it navigates this landscape, the focus shifts to the intricate components of electrified vehicles, including the electric motor/generator, batteries and their management system, supercapacitors, onboard rechargers, high-voltage harness, high-voltage electronic modules, fuel cells, and hydrogen tanks [7.8, 7.9].

These components work harmoniously to manage these advanced vehicles' complex energy storage, distribution, and conversion interplay. The exploration extends to the safety features encompassing manual service disconnects, high-voltage disconnection systems, and innovative regenerative braking technology. Finally, it highlights the importance of recharging stations, their connectors, and the emerging technology of inductive rechargers, emphasizing the evolving infrastructure that underpins the widespread adoption of electrified vehicles.

7.2.
Configurations and Acronyms

After over a century of vehicular reliance on fossil fuels, notably gasoline and diesel, shifting environmental priorities and advancing technologies are reshaping the landscape of energy sources in the automotive industry. With growing emphasis on environmental sustainability and corresponding regulations, we face a more diverse array of power sources. Some proponents of this shift envision a future where "pure" EVs predominate, supplanting traditional ICEVs. However, along this transition, many acronyms have emerged to describe the energy sources and powertrain configurations.

Adding a layer of complexity, the interpretation of these acronyms may vary depending on the country, manufacturer, local regulations, and context. This section aims to unravel the intricacies of various vehicle configurations and essential acronyms associated with this transformative phase. By doing so, it seeks to clarify the distinct components defining these cutting-edge vehicular systems.

7.2.1.
BEV

BEVs stand out as a paradigm shift within electrified powertrain vehicles by relying solely on an electric traction motor for propulsion without incorporating an ICE. Central to this design is a substantial traction battery, accompanied by sophisticated control modules and occasionally referred to as "pure EVs," BEVs are sometimes simplified to the abbreviation EV, a term embraced by the general public and nontechnical media, albeit with the potential for confusion with other electrified powertrain configurations. **Figure 7.1** serves as a visual guide, elucidating the primary components inherent to BEVs. Specific models exhibit a dual-electric motor setup, while others showcase a unique design featuring a dedicated electric motor within each wheel.

Figure 7.1 Main components of a BEV.

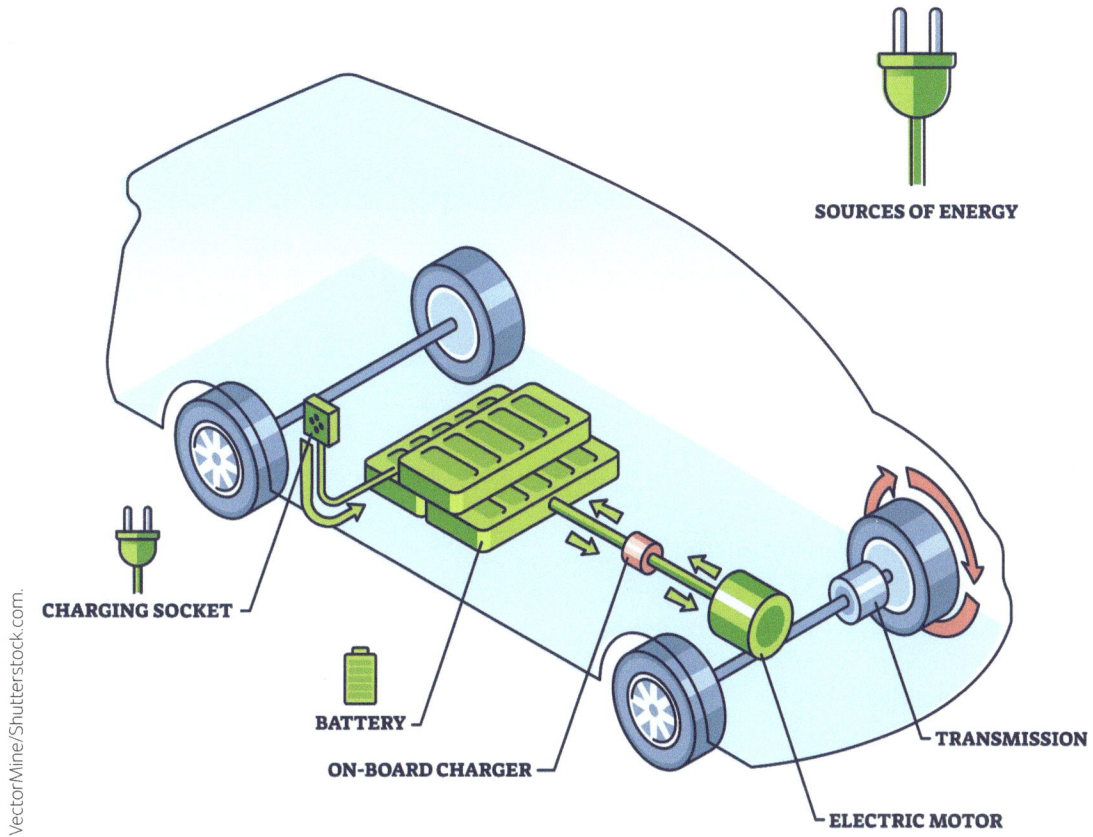

BEV
BATTERY ELECTRIC VEHICLE

SOURCES OF ENERGY

CHARGING SOCKET

BATTERY

ON-BOARD CHARGER

TRANSMISSION

ELECTRIC MOTOR

VectorMine/Shutterstock.com.

7.2.2.
EV

The term "EV," denoting electric vehicle, carries a spectrum of meanings influenced by factors such as geographical location, regulatory frameworks, usage classification, and the context in which it is employed, including discussions within the "general public" and "non-technical" forums. In a broad sense, an EV is characterized by an electrified powertrain featuring an electric traction motor, a traction battery, and associated control modules. However, it is crucial to recognize that regional variations may lead to instances where an EV incorporates an ICE or a fuel cell.

This book intentionally refrains from using the acronym EV to enhance clarity and precision in communication. Instead, the term BEV is employed to specifically refer to "pure" EVs, signifying those relying solely on electric traction motors without ICEs. The abbreviation xEV is adopted to encompass vehicles with an electrified powertrain, with or without an ICE, and with or without a fuel cell. This distinction aims to mitigate potential ambiguity and ensure consistent terminology throughout the discussion of different configurations of vehicles with an electrified powertrain.

7.2.3.
FCEV or FCV

Both abbreviations are used to denote fuel cell EVs. The core of this technology is the fuel cell, and these vehicles typically feature a sizable hydrogen tank. Hydrogen undergoes a transformative process within the fuel cell, diverging from traditional combustion and combining with atmospheric oxygen to release electricity and generate water vapor. This electricity powers the electric traction motor, with any surplus or energy reclaimed from regenerative braking stored in a smaller battery (compared to the battery used in BEVs). This innovative approach propels the vehicle and underscores a sustainable path, emitting only water vapor as a by-product.

7.2.4.
HEV

According to SAE Standard J1715, an HEV is a vehicle that can draw propulsion energy from both of the following sources of stored energy: (1) a consumable fuel and (2) a rechargeable energy storage system (RESS) that is charged by an electric motor–generator system, an external electric energy source, or both. The most common RESS is an extensive LIB system, but other mechanisms exist, such as high-speed flywheels.

In other words, the HEV category encompasses vehicles featuring a consumable fuel energy source and an electrical energy source for the powertrain, as illustrated in **Figure 7.2**. Consumable fuels vary and include gasoline, diesel, alcohol, and LPG.

The classification of HEVs can be further refined on subcategories based on its primary components' relative size, arrangement, and intended mission. These distinct HEV subcategories provide insight into the diverse landscape within the hybrid vehicle class as discussed in the following sections.

Figure 7.2 Main components of an HEV.

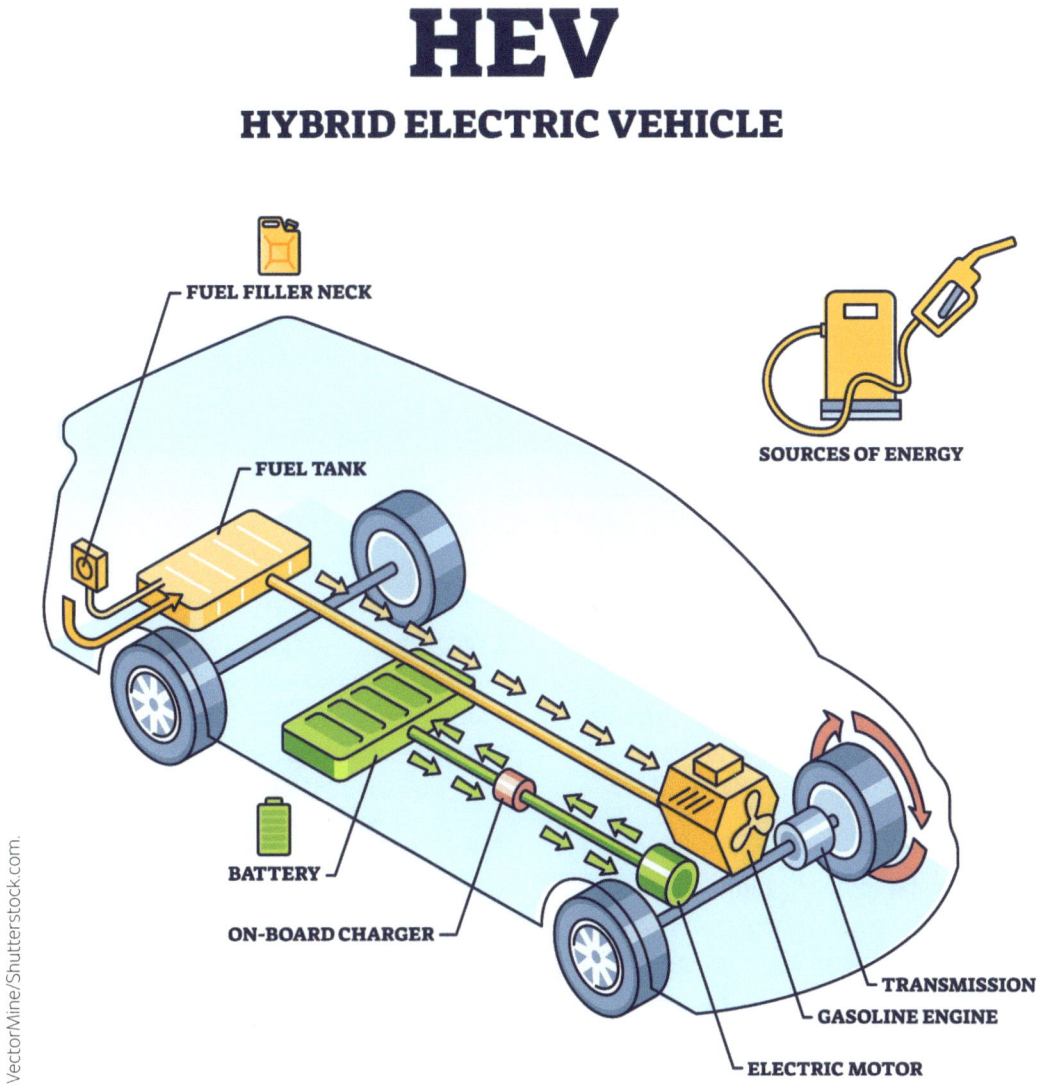

HEV
HYBRID ELECTRIC VEHICLE

FUEL FILLER NECK

FUEL TANK

SOURCES OF ENERGY

BATTERY

ON-BOARD CHARGER

TRANSMISSION

GASOLINE ENGINE

ELECTRIC MOTOR

7.2.4.1.
Micro HEV (MHEV)

MHEVs, also known as start–stop hybrids, use a start–stop system to automatically shut off the engine when the vehicle stops, such as at traffic lights or during idling. The engine restarts promptly when the driver presses the accelerator or engages the clutch. This technology helps conserve fuel and reduce emissions but does not provide electric-only propulsion.

7.2.4.2.
Mild HV (MHV or mHEV)

MHVs typically have a more advanced electrification system than micro hybrids. They usually incorporate a small electric motor, often known as an electric assist motor, which collaborates with the ICE during acceleration, deceleration, or cruising. MHVs can also recover and store energy during braking, improving fuel efficiency. While they offer more electrification features than micro hybrids, they do not usually provide sustained electric-only driving and are considered less electrified than full hybrids.

A caution: Micro and Mild hybrid terminologies are sometimes used interchangeably. It is essential to consider specific features and technologies different manufacturers implement when analyzing micro and mild hybrid vehicles.

7.2.4.3.
(Full) HEV

HEV stands for hybrid electric vehicle, and usually this acronym refers to a "full" hybrid, in contrast with the MHEV and MHV configurations. A full HEV features a substantial electric motor and a sizable battery, enabling the vehicle to cover a significant range without activating its ICE. The ICE comes into play during lengthier journeys, when there is a need to recharge the traction battery, or when the electric motor alone cannot deliver the required acceleration.

7.2.4.4.
Plug-In Hybrid Electric Vehicle (PHEV)

PHEV is elucidated in Figure 7.3, showcasing its key components. A PHEV introduces a novel facet unlike a conventional HEV, which solely replenishes its battery using energy from the ICE or recovered from regenerative braking. A PHEV traction battery can also be recharged from an external electric source, such as a grid recharger. This distinctive feature aims to transform the vehicle into a BEV when desired.

However, this transition to plug-in capability comes with trade-offs, including heightened complexity and cost. A PHEV's electric-only range is typically more limited than an equivalent BEV. Balancing these considerations, PHEVs offer a versatile solution catering to both "pure" electric and "conventional" hybrid driving scenarios.

Figure 7.3 Main components of a PHEV.

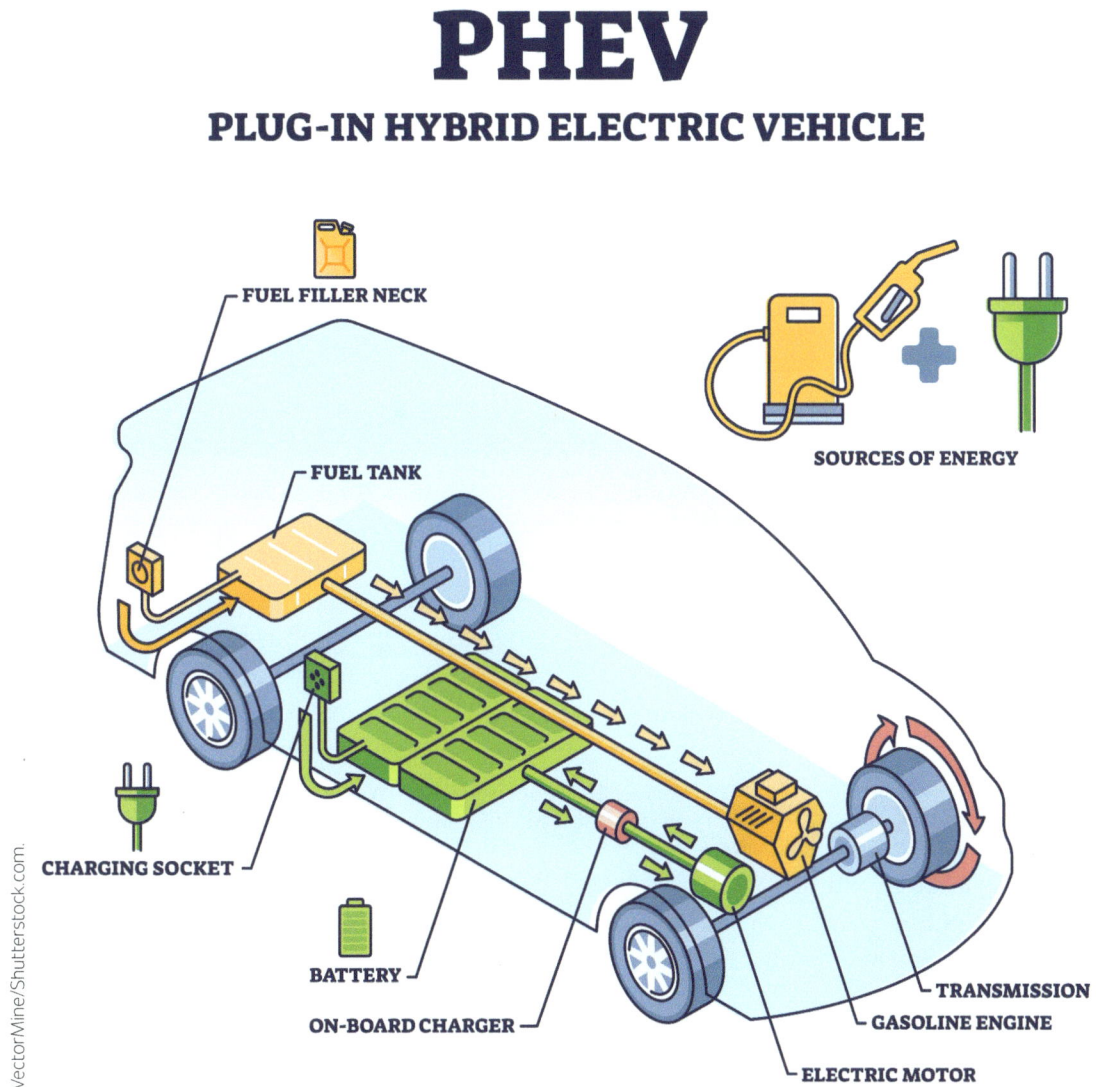

PHEV
PLUG-IN HYBRID ELECTRIC VEHICLE

FUEL FILLER NECK

SOURCES OF ENERGY

FUEL TANK

CHARGING SOCKET

BATTERY

ON-BOARD CHARGER

TRANSMISSION

GASOLINE ENGINE

ELECTRIC MOTOR

VectorMine/Shutterstock.com.

7.2.4.5.
Series and Parallel Hybrids

This classification, pertinent to HEV and PHEV categories, intricately examines the primary energy pathways. The ICE is divorced from directly propelling the transmission in series hybrids. Instead, it energizes a generator, and the generated output powers the electric traction motor. Conversely, parallel hybrids simultaneously enable the thermal engine and the electric motor to engage the transmission or drive the wheels actively.

In both series and parallel configurations, the intricacies of regenerative braking and surplus energy mechanisms come into play, facilitating the recharging of batteries. This nuanced distinction in energy flow highlights the diverse approaches within the hybrid landscape, tailoring solutions to specific efficiency and performance objectives.

7.2.5.
ICE

ICE is not only an acronym for the engine technology but is sometimes used to refer to vehicles utilizing an ICE in its powertrain. Across most nations, gasoline engines dominate the landscape of lightweight and highway passenger vehicles, while diesel engines traditionally power larger trucks and buses.

In recent decades, certain countries have diversified their fuel sources, incorporating alternatives such as ethanol, methanol, LPG, CNG, and biodiesel (derived, in part, from plantations). This evolution extends to "flex" engines capable of utilizing at least two different fuels, offering versatility with options like gasoline and ethanol.

Regardless of the fuel source, ICEs burn a combustible fluid, propel pistons, and generate mechanical energy to drive the vehicle's wheels. Unfortunately, this combustion process generates tailpipe emissions and greenhouse gases, prompting ongoing efforts to explore environmentally conscious alternatives. For instance, initiatives exploring hydrogen as a fuel for ICEs are underway, aiming for nearly emission-free operation, with only water vapor and limited NOx emissions.

7.2.6.
ICV or ICEV

While not frequently used, both ICV and ICEV abbreviations stand for internal combustion engine vehicle. These acronyms align with the conventions employed for subsequent discussions on electrified powertrains. ICV and ICEV collectively denote vehicles with a conventional combustion engine powered by fossil fuel or alcohol, lacking an electric traction motor or traction battery. This category represents the predominant type within the current global vehicle fleet.

In an era where the automotive industry is transitioning toward electrified powertrains, the distinction provided by the ICV/ICEV label becomes more relevant. While this type of vehicle traditionally dominated car markets, the differentiation was primarily based on the fuel type (i.e., gasoline or diesel). However, along this transition, the ICV/ICEV label becomes a valuable identifier in discussions about the evolving landscape of vehicle propulsion technologies.

7.2.7.
New Energy Vehicles (NEV)

The acronym NEV commonly refers to "new energy vehicles." Although the term is not exclusive to any particular country, it is often utilized in the context of Chinese regulations and policies related to electric and plug-in hybrid vehicles. The designation "NEV" in China encompasses various electrified vehicles, including BEVs, PHEVs, and FCEVs.

7.2.8.
Plug-In Electric Vehicle (PEV)

PEVs refers to electrified vehicles that can be recharged with an off-board source of electricity; it includes both BEVs and PHEVs.

7.2.9.
xEV

xEV refers to any electrified propulsion vehicle with a high-voltage system, encompassing, but not limited to, BEVs, HEVs, PEVs, and FCEVs. Using the abbreviation xEV serves to disambiguate, particularly in contexts or countries where EVs refer to "pure" BEVs.

7.2.10.
Zero-Emission Vehicle (ZEV)

ZEV stands for zero-emission vehicle, denoting a vehicle that generates no tailpipe emissions. These vehicles are designed to minimize their environmental impact and decrease air pollution. ZEVs encompass various types of xEVs, including BEVs and FCEVs that operate using hydrogen. In some regulatory frameworks, PHEVs may be considered ZEVs when operating in electric-only mode. ZEVs play a crucial role in reducing greenhouse gas emissions and diminishing reliance on traditional ICEVs, thereby contributing to efforts to mitigate the environmental impact associated with transportation.

7.3.
Components

This section places a spotlight on distinctive components integral to electrified vehicles. Notably, the examination centers on components unique to these vehicles, as potential fire risks associated with components inherited from ICEVs have been thoroughly explored in preceding chapters.

7.3.1.
Electric Motor/Generator

Serving as the electric counterpart to the thermal engine in ICEVs, the electric motor assumes a pivotal role in xEVs. Fed by electric energy via a power control module, the electric motor creates the rotational movement essential for propelling the wheels. In specific configurations, a singular motor drives a straightforward transmission or more intricate versions, some even incorporating automatic gears/transmission, with the transmission subsequently propelling the wheel shafts. Alternatively, some vehicles integrate multiple motors, possibly having one motor per wheel, obviating the need for a transmission.

During braking maneuvers, the electric motor seamlessly transitions into a generator, harnessing kinetic energy from the wheels and converting it into electricity. This electricity, in turn, replenishes the battery while the vehicle decelerates—a process commonly referred to as regenerative braking.

7.3.2.
Low-Voltage Battery

While high-voltage LIB packs propel xEVs into the future, low-voltage batteries, often 12 V lead–acid, maintain their significance. Traditionally entrenched in ICEVs to power auxiliary components like lights, power windows, and electronic control units, low-voltage batteries persist as crucial elements in the evolving automotive landscape.

Within the realm of xEVs, these low-voltage batteries assume a dual role. Beyond serving as essential backup power for safety systems, they continue to power ancillary functions such as infotainment and security systems designed for a 12 V electrical system. Integrating low-voltage batteries aligns with the automotive industry's commitment to a well-established 12 V infrastructure. This strategic alignment ensures compatibility, capitalizing on proven technologies and streamlining integration efforts.

Moreover, the deliberate separation of low-voltage systems enhances safety by allowing the high-voltage battery to focus exclusively on propelling the vehicle. Low-voltage batteries in xEVs uphold the reliability of time-tested technologies and enable seamless integration into the established automotive infrastructure.

7.3.3.
Traction Battery

Traction batteries are the fundamental energy source propelling xEVs when operating in electric mode. The era of mass-producing HEVs began in 1997 with the introduction of the Toyota Prius in Japan, featuring nickel–metal hydride (NMH) batteries—a technology swiftly adopted by other manufacturers for their inaugural HEV models. In subsequent years, the industry experienced a transition toward lithium-ion alternatives, thanks to their higher energy density, primarily in BEVs. In 2012, most carmakers (except Toyota) switched their HEVs to LIBs.

Referred to as LIB, these advanced batteries play a pivotal role in storing electrical energy, influencing the driving range, and determining the overall performance of xEVs. The energy capacity of traction batteries far exceeds that of traditional 12 V batteries in ICEVs by several orders of magnitude.

To understand the historical dominance of ICEVs, a comparison of the energy density between gasoline, the traditional fuel for ICEVs, and BEV batteries is essential. Gasoline exhibits high energy density, approximately 115,000 BTU/gal (around 34 MJ/L) or about 18,000 BTU/lb (approximately 45 MJ/kg), making it highly suitable for combustion engines. In contrast, traditional lead–acid batteries used in ICEVs historically have much lower energy densities: 450 BTU/gal (about 0.13 MJ/L) and 54 BTU/lb (approximately 0.13 MJ/kg). Achieving a comparable driving range to current ICEVs with lead–acid batteries is impractical, as they would necessitate batteries occupying a volume and weight several orders of magnitude larger than the typical ICEV gasoline tank.

The viability of electrified vehicles as an alternative hinges on advancements in battery technologies. Modern xEVs commonly employ

LIBs, offering significantly higher energy densities—approximately 2600 BTU/gal (around 0.72 MJ/L) and 200 BTU/lb (approximately 0.47 MJ/kg). Ongoing research and development aims to enhance these figures through improved battery chemistry, materials, and manufacturing processes. **Figure 7.4** presents a Ragone plot comparing various automotive battery technologies' power and energy densities, illustrating progress and technological alternatives [7.10-7.16].

Figure 7.4 Ragone plot.

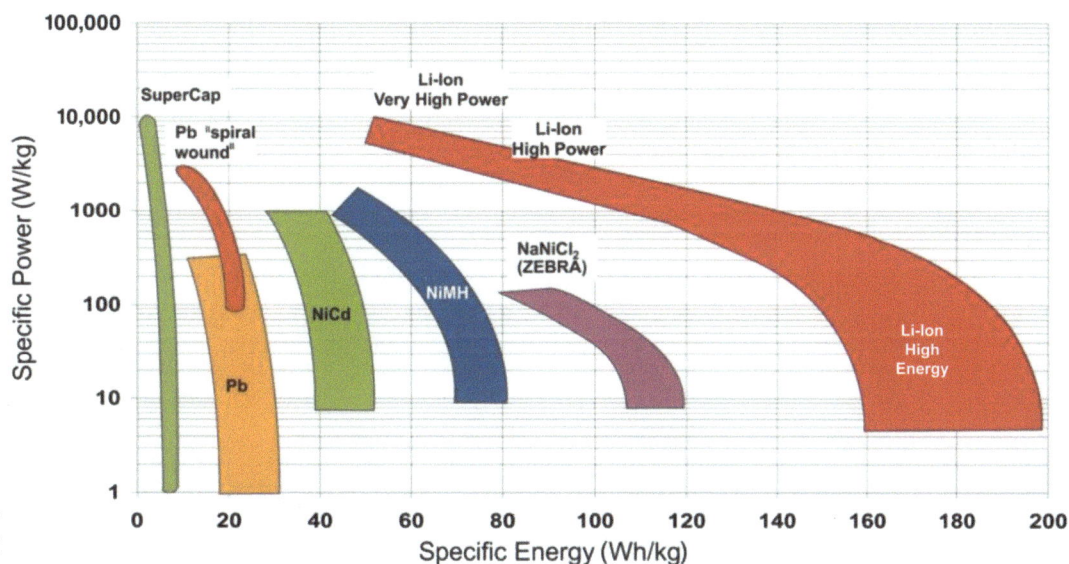

Budde-Meiwes, H.; Drilkens, J.; Lunz, B.; Muennix, J.; Rothgang, S.; Kowal, J.; Sauer, D. A review of current automotive battery technology and future prospects. J. Automob. Eng. 2013, 227, 761–776, permission granted by Sage Publications through Copyright Clearance Center.

While individual cells within these batteries exhibit relatively low voltage and energy storage capabilities, combining hundreds or thousands of them in series and parallel connections and in battery submodules allows the complete battery system to achieve voltages typically between 200 and 800 V, storing between 50 and 100 kWh.

Figure 7.5 illustrates some of the diverse shapes of individual cells, submodules, and battery systems used by different manufacturers.

The relevant failure mechanisms of lithium batteries will be examined in Chapter 8.

Figure 7.5 Typical xEV batteries.

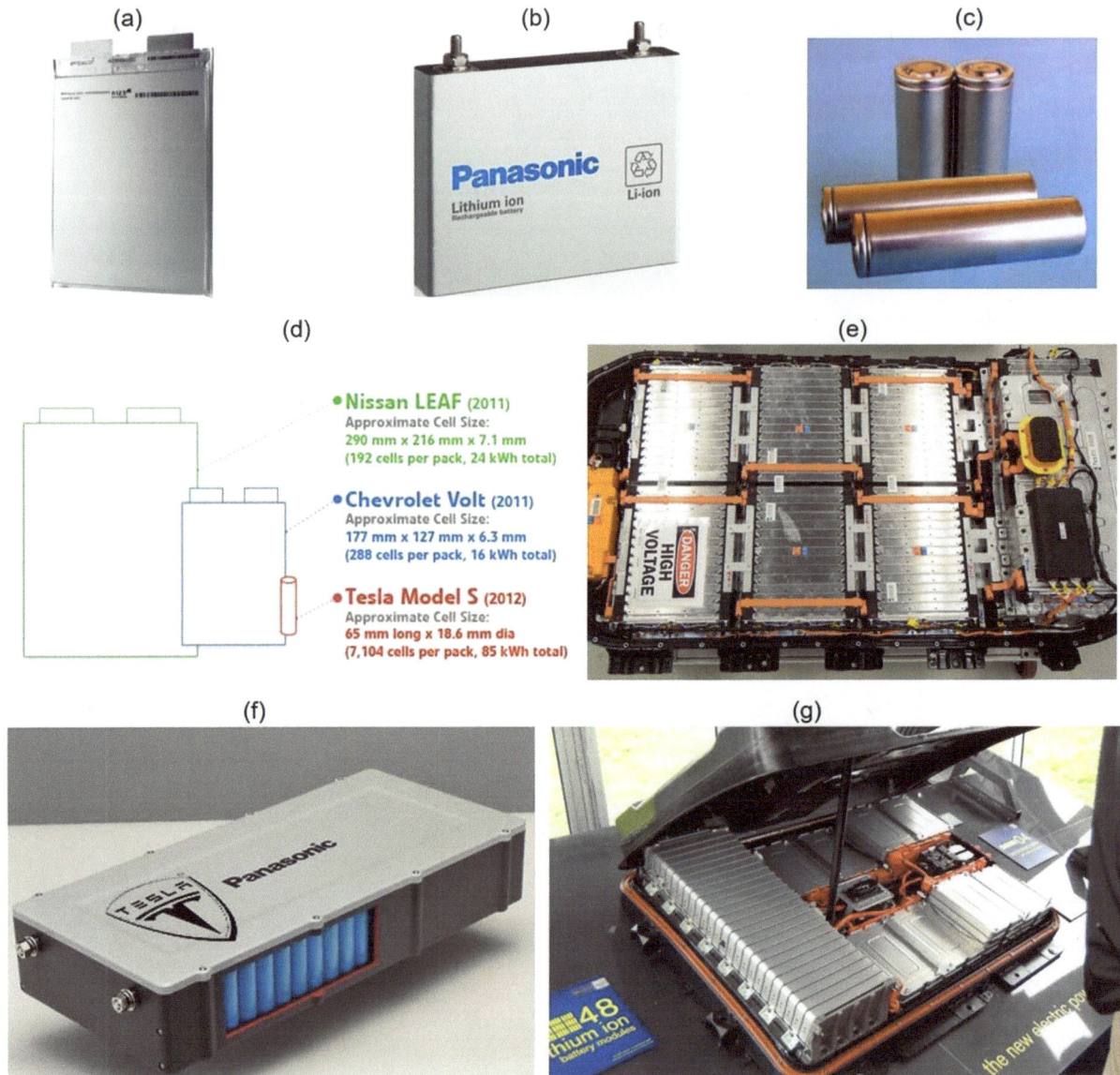

(a)

(b)

(c)

(d)

- **Nissan LEAF** (2011)
  Approximate Cell Size:
  290 mm x 216 mm x 7.1 mm
  (192 cells per pack, 24 kWh total)

- **Chevrolet Volt** (2011)
  Approximate Cell Size:
  177 mm x 127 mm x 6.3 mm
  (288 cells per pack, 16 kWh total)

- **Tesla Model S** (2012)
  Approximate Cell Size:
  65 mm long x 18.6 mm dia
  (7,104 cells per pack, 85 kWh total)

(e)

(f)

(g)

Sun, Peiyi & Bisschop, Roeland & Niu, Huichang & Huang, Xinyan. (2020). A Review of Battery Fires in Electric Vehicles. Fire Technology. 1-50. 10.1007/s10694-019-00944-3, permission granted by Springer Nature through Copyright Clearance Center.

7.3.4.
Battery Management System (BMS)

The BMS is a pivotal component in xEVs. It goes beyond mere monitoring, actively overseeing the traction battery's health, safety, and performance. Essential functions include managing the state of charge (SOC) and state of health (SOH), ensuring optimal temperature levels through distributed sensors and a cooling system, and monitoring voltage and current across the battery [7.17-7.19].

The BMS plays a crucial role in maintaining cell balance within multicell battery packs, thereby averting performance degradation and potential damage. Embedded safety features, encompassing overcurrent and overvoltage protection and under-temperature and overtemperature protection, contribute to shielding against possible hazards. Additionally, the BMS is equipped to detect loss of insulation and loss of communication, ensuring a comprehensive safety net. Should abnormal conditions arise, the BMS can take proactive measures, including disconnecting battery submodules or executing an emergency shutdown.

Moreover, the BMS acts as a communication hub, interacting with various modules and components in the xEV and with external rechargers. Collaboration with the electric motor's power control module manages power delivery, while coordination with the onboard charger optimizes charging. Additionally, the BMS works with the regenerative braking control to adequately capture and store energy during braking.

Beyond its real-time monitoring capabilities, the BMS logs data related to battery performance, faults, and environmental conditions. This comprehensive data logging facilitates diagnostics, maintenance, and continuous improvement of battery technology. The BMS is a sophisticated controller that ensures the traction battery's efficiency, safety, and longevity in xEVs.

7.3.5.
Supercapacitors

Supercapacitors, also known as ultracapacitors or electrochemical capacitors, are electronic devices capable of storing large amounts of electrical charge in a much smaller volume than typical dielectric capacitors. In contrast to dielectric capacitors, which store charges in metallic plates surrounding a dielectric foil, supercapacitors utilize an ultrathin dielectric layer on the surface of an electrolyte, measured in nanometers (nm) or even angstroms (Å), enabling the accumulation of electrostatic charges through ions.

Another notable advantage of supercapacitors is their efficiency in handling high bursts of power during charging and discharging. This characteristic makes them particularly valuable in regenerative braking systems, relieving the traction battery of stress. Despite managing significant power peaks, the service life of ultracapacitors typically exceeds that of lithium batteries, contributing to enhanced battery lifespan, improved efficiency, and extended vehicle range.

However, it is essential to acknowledge their limitations, such as lower energy density compared to batteries, besides increasing the cost and complexity of the vehicle. They operate under relatively small voltages, necessitating the combination of multiple units to match the high voltages in xEVs. Additionally, specific risks are associated with their capability to retain electric energy for extended periods, even with the vehicle turned off, and risks with some potentially generating flammable and toxic substances, such as methyl cyanide.

Ongoing research and development efforts aim to address these challenges, seeking to broaden the applicability of supercapacitors in the automotive industry. Concurrently, advancements in traction batteries are enhancing tolerance and efficiency during charging and discharging bursts, potentially eliminating the need for supercapacitors.

7.3.6.
Onboard Recharger

The onboard recharger, often referred to as an onboard charger (OBC), is a crucial component responsible for converting electricity from an external power source, such as a wall outlet or a public charging station, into the appropriate level of direct current (DC) electricity to recharge the traction battery. Certain PEVs require the external source to be alternating current (AC); others require a DC source, making the OBC a vital component enabling these vehicles to replenish their batteries adequately.

The OBC plays a distinct but interconnected role during regenerative braking by directing recovered electricity to the traction battery for storage. In specific xEV configurations, supercapacitors may complement the traction battery. In these setups, the onboard recharger orchestrates the energy split, channeling rapid bursts of power to the supercapacitors and allocating a less demanding portion for storage in the traction battery. This strategic allocation ensures efficient utilization of energy resources and optimizes the overall performance and service life of the vehicle energy storage system.

7.3.7.
High-Voltage Harness

Within xEVs, the high-voltage harness comprises a network of cables and wiring designed to convey electrical power generated by the high-voltage battery to diverse components within the vehicle. Operating at several hundred volts, the high-voltage system powers critical components integral to the vehicle's propulsion, including the electric traction motor and power electronics.

Orange is widely used for high-voltage components and harnesses, as exhibited in **Figure 7.6**, to boost safety. This standardized identification, adhering to SAE J1673, aids technicians, first responders, and maintenance personnel in quickly recognizing high-voltage elements. The vivid orange hue is a visual cue that highlights dangerous components from other electrical systems and prompts necessary precautions [7.20].

Figure 7.6 Orange high-voltage harness.

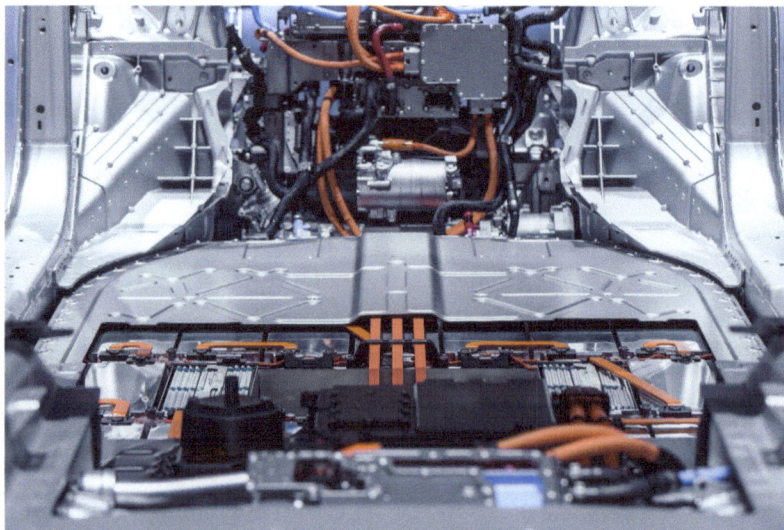

Sergii Chernov/Shutterstock.com.

7.3.8.
High-Voltage Electronic Modules

In addition to overseeing the energy supply for the electric motor, specialized modules play a crucial role in facilitating various functions in xEVs. These functions encompass the onboard battery recharger, regenerative braking recovery system, battery management system, low-voltage converter, and more. Integrating some or all of these functions into single or multiple modules is common in contemporary xEVs. Furthermore, certain vehicles incorporate high-power systems such as electric steering and adaptive suspension systems, drawing power directly from the high-voltage distribution system.

7.3.9.
Manual Service Disconnect (MSD)

The MSD, also known as the service disconnect plug, manual disconnect, or high-voltage disconnect, is a crucial safety feature designed to physically separate a vehicle's high-voltage electrical system from its power source. Serving as a manual switch or plug, it enables technicians, emergency responders, and those working on the vehicle to disconnect the high-voltage battery pack during maintenance or emergencies, ensuring their safety by preventing electric shock or exposure to high-voltage components. Mandated by regulations and standards, this mechanism allows rapid de-energization of the high-voltage system in case of accidents, minimizing electrical hazards and protecting service personnel and emergency responders interacting with xEVs. Figure 7.7 shows one example of such a device [7.21, 7.22].

Figure 7.7 Mini MSD (image Courtesy of Amphenol PCD).

Courtsey of Amphenol.

7.3.10.
High-Voltage Disconnection

Besides the MSD, *automatic* high-voltage disconnection systems in xEVs prioritize safety throughout regular vehicle operation and in various scenarios, such as during maintenance, emergencies, or collisions. The high voltage system's positive and negative poles are engaged only when the driver turns on the vehicle. Yet, several safety mechanisms can disconnect the high voltage, activated by events like airbag deployment, collision or turn over detection, detection of leakage current from the high-voltage battery, thermal protection, and loss of communication with vital modules. Notably, in some xEVs, pyrotechnic fuses serve as a specific mechanism to rapidly and reliably disconnect the energy in emergencies, as exemplified in Figure 7.8. A small explosive load in this device literally cuts the electric connection with the high-voltage battery. These measures collectively ensure the safety of individuals interacting with electric and hybrid vehicles.

Figure 7.8 Pyrotechnic high-voltage fuse.

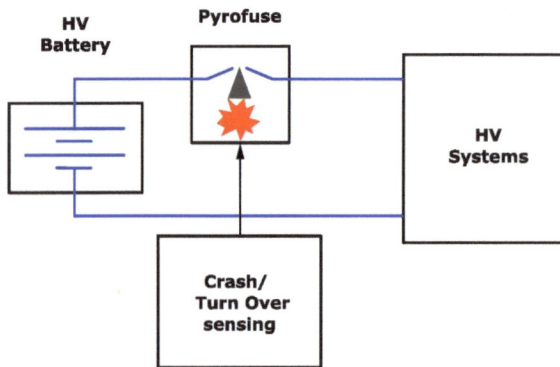

7.3.11.
Fuel Cells

Fuel cells generate electrical energy through an electrochemical reaction, usually between hydrogen and oxygen. In the operation of a hydrogen fuel cell, hydrogen gas is supplied to the anode, where it undergoes a process of splitting into protons and electrons. Protons move through the electrolyte (a proton exchange membrane, in one of the possible technologies), while electrons create an electric current in the external circuit. At the cathode, oxygen from the air combines with protons and electrons to form water, producing electricity. This electricity is then used to power an electric motor, propelling the vehicle forward. FCEVs store hydrogen in onboard tanks; water vapor is the only emission from the tailpipe.

Fuel cell technology offers advantages such as higher efficiency, more extended range, and shorter refueling times than certain BEVs. It provides a clean and sustainable option for

zero-emission transportation, mainly when hydrogen is produced using renewable energy sources. However, challenges such as developing a hydrogen infrastructure, the energy-intensive process of hydrogen production, and the overall cost of fuel cell systems are areas of ongoing research and development within the automotive industry.

7.3.12.
Hydrogen Tank

In the operational dynamics of an FCEV, the hydrogen tank assumes a crucial role by serving as the repository for compressed hydrogen gas. Its principal responsibility is to securely store hydrogen, a lightweight gas that necessitates compression to high pressures ranging from 350 to 700 bar (5000 to 10,000 psi), to provide the necessary driving range. Even at this pressure, hydrogen remains in the gaseous state, which implies in low density and large tanks, made of advanced materials.

Ensuring the purity of stored hydrogen, minimizing storage losses over time, and enhancing overall storage efficiency and safety are additional challenges that researchers and engineers grapple with. Furthermore, issues related to establishing a widespread hydrogen refueling infrastructure and standardizing protocols require careful attention. Addressing these challenges is vital for advancing the adoption of FCEVs, and ongoing research endeavors focus on improving the safety, efficiency, and cost-effectiveness of hydrogen storage technologies.

7.3.13.
Regenerative Braking

Regenerative braking is not a physical component but a technology prevalent in electrified vehicles. It is designed to recover and store kinetic energy, typically as electric energy, during braking. In traditional ICEVs, when the brakes are applied, kinetic energy is converted into heat, wearing out the brake pads and reducing the vehicle's speed. Applying the brakes in electrified vehicles equipped with regenerative braking transitions the electric motor into generator mode, converting kinetic energy into electrical energy.

The recaptured electrical energy is stored in the vehicle's storage system for later use, particularly during acceleration or heightened power demand, thereby extending the vehicle's range. Technical constraints exist regarding the amount of kinetic energy that can be converted into electricity, influenced by vehicle speed, intended deceleration, and the type of electric storage systems onboard. Consequently, electrified vehicles still incorporate friction brakes within their braking systems, but these handle a smaller portion of the braking task.

7.4.
Recharging Stations and Connectors

Recharging stations, also known as charging stations or charging points, are facilities designed to supply electric power to recharge the batteries of PEVs. They are typically equipped with one or more charging points connected to the power grid. Charging stations can be found in public places such as shopping centers, parking lots, and rest areas, as well as in private homes and businesses. For safety reasons, in line with SAE J1772, PEVs prevent operator-intended vehicle movement when the recharging connector is mated to the vehicle inlet.

However, the proliferation of incompatible connector standards has hindered the global implementation of charging infrastructure. Beyond mechanical variations in connectors, some vehicles accept only external AC energy, others DC, and some at specific voltage levels, increasing compatibility issues [7.23-7.25].

Some connectors in use include the following types:

- Type 1 (SAE J1772): Common in North America and some parts of Asia.
- Type 2 (IEC 62196): Widely used in Europe.
- Combined charging system (CCS): This is an extension of the Type 2 connector with additional pins for high-power DC charging. It is used in Europe and North America.
- CHAdeMO: Primarily used by Japanese and some European and American manufacturers for high-power DC charging.
- GB/T: Used in China.
- NACS, also known as SAE J3400, and Tesla Supercharger Connector. It might become the de facto standard for North America.

The North American Charging Standard (NACS) is a PEV charging connector system that Tesla, Inc. developed. It has been used on all North American market Tesla vehicles since 2012, was opened for use to other makers in November 2022, and standardized as SAE J3400 in December 2023. The NACS connector is a unified connector for AC and DC recharging, capable of handling up to 350 kW and supporting all voltage levels. Its recent adoption by major automakers in the US will hopefully accelerate the implementation of a charging infrastructure compatible with any PEV vehicle, regardless of its manufacturer.

7.5.
Inductive Rechargers

Inductive rechargers are wireless charging technology that transfers electric energy from a charging pad or coil to the vehicle without needing a physical connection. This technology is based on electromagnetic induction, where an AC in a charging coil generates a magnetic field, and a corresponding coil in the vehicle converts this magnetic field back into electric current to charge the battery [7.26].

The following sections discuss the two main types of inductive rechargers for PEVs.

7.5.1.
Inductive Rechargers without a Paddle

In systems without a paddle, as exemplified in **Figure 7.9**, the PEV is equipped with a receiving coil, typically located on the vehicle's underside. When the PEV is parked over the inductive charging pad on the ground, coils of the charging pad and those of the vehicle align, and the charging process begins. This type of inductive charging is convenient as it eliminates the need for a physical cable or plugs. However, precise alignment between the charging pad and the vehicle is crucial for efficient charging. Ongoing research explores extending inductive recharging to electric buses at passenger stations and vehicles on highways, even when in motion, despite reduced efficiency.

Figure 7.9 Inductive recharging.

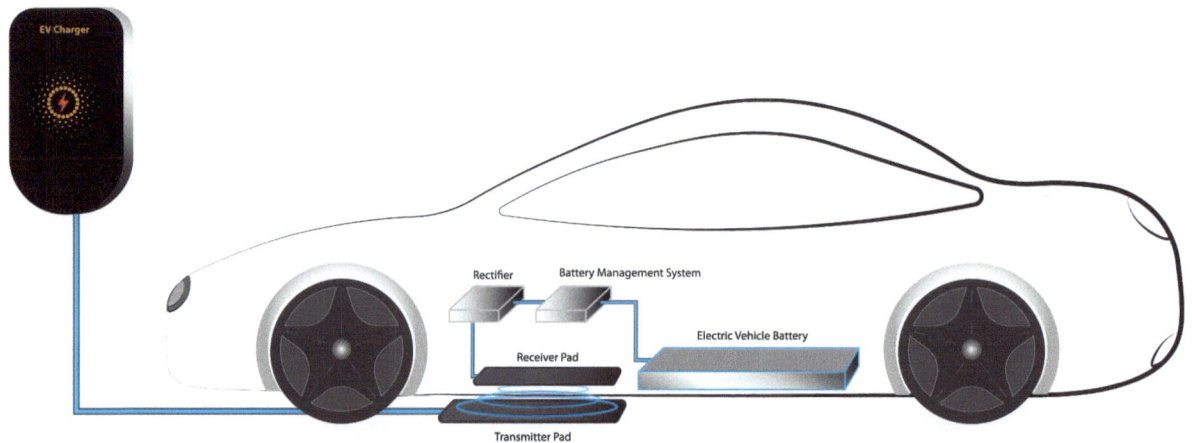

7.5.2.
Inductive Rechargers with a Paddle

Some inductive charging systems for early PEVs used a paddle, as defined in SAE J1773, placed in a corresponding inlet in the vehicle. This paddle ensured accurate alignment, making the inductive charging process more reliable and efficient. However, it had limited power transfer compared to wired connections. **Figure 7.10** illustrates one of these paddles manufactured by Hughes a few decades ago, which has since been discontinued.

Figure 7.10 Inductive paddle.

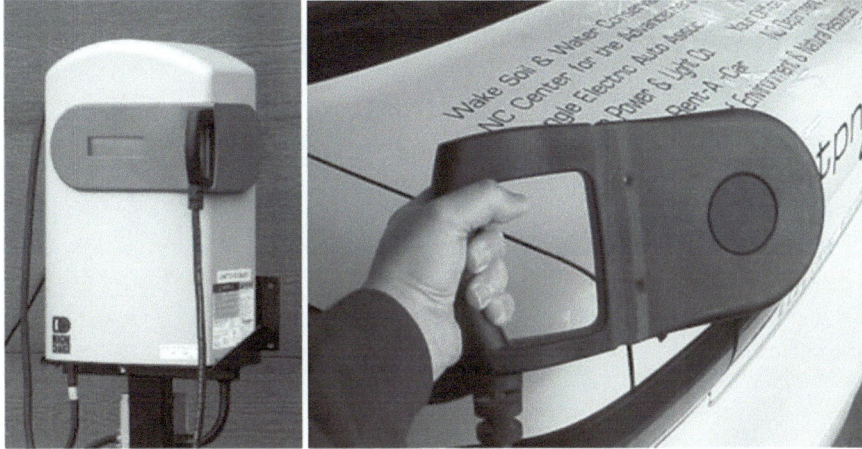

Musavi, F. and Eberle, W. (2014), Overview of wireless power transfer technologies for electric vehicle battery charging. IET Power Electronics, 7: 60-66. https://doi.org/10.1049/iet-pel.2013.0047, permission granted by John Wiley and Sons through Copyright Clearance Center.

Inductive charging offers the convenience of connector or cable-free usage, mainly when plugging in a cable might be impractical. However, it is less efficient than wired charging methods, and the infrastructure for inductive charging is incipient. With technological advancements, inductive charging systems may become more widespread and integrated into various environments to enhance the charging experience for xEV users.

References

[7.1].   Johnston, C. and Sobey, E., *The Arrival of the Electric Car*, SAE International Book R-534 (Warrendale: SAE International, 2023), ISBN:978-1-4686-0502-0.

[7.2].   Kershaw, J., *SAE International's Dictionary for Automotive Engineers*, SAE International Book R-523 (Warrendale: SAE International, 2023), ISBN:978-1-4686-0407-8.

[7.3].   SAE International Ground Vehicle Standard, "Hybrid Electric Vehicle (HEV) and Electric Vehicle (EV) Terminology," SAE Standard J1715, J1715_202209, September 2022.

[7.4].   Senecal, K. and Leach, F., *Racing Toward Zero: The Untold Story of Driving Green*, SAE International Book R-501 (Warrendale: SAE International, 2021), ISBN:978-1-4686-0146-6.

[7.5].   National Archives, "United States Code of Federal Regulations, 40 CFR Part 86, Control of Emissions from New and In-Use Highway Vehicles and Engines," accessed December 2023, https://www.ecfr.gov/current/title-40/part-86.

[7.6].   National Archives, "United States Code of Federal Regulations 40 CFR Part 600," accessed December 2023, https://www.ecfr.gov/current/title-40/part-600.

[7.7].   Wakefield, E., *History of the Electric Automobile*, SAE International Book R-187 (Warrendale: SAE International, 1998), ISBN:978-0-7680-3749-4.

[7.8].   Alam, M., Ahmad, A., Khan, Z., Rafat, Y. et al., "A Bibliographical Review of Electrical Vehicles (xEVs) Standards," *SAE Int. J. Alt. Power.* 7, no. 1 (2018): 63-98, doi:https://doi.org/10.4271/08-07-01-0005.

[7.9].   Kalaskar, V., Conway, G., Handa, G., Joo, S. et al., "Challenges and Opportunities with Direct-Injection Hydrogen Engines," SAE Technical Paper 2023-01-0287 (2023), doi:https://doi.org/10.4271/2023-01-0287.

[7.10]. Avicenne, "The Rechargeable Battery Market and Main Trends 2018-2030," ICBR 2019, accessed January 2024, https://rechargebatteries.org/wp-content/uploads/2019/02/Keynote_2_AVICENNE_Christophe-Pillot.pdf.

[7.11].   Budde-Meiwes, H., Drillkens, J., Lunz, B. et al., "A Review of Current Automotive Battery Technology and Future Prospects," *Proceedings of the Institution of Mechanical Engineers, Part D: Journal of Automobile Engineering* 227 (2013): 761-776, doi:https://doi.org/10.1177/0954407013485567.

[7.12].   SAE International Ground Vehicle Standard, "Storage Batteries," SAE Standard J537, J537_202309, September 2023.

[7.13].   SAE International Ground Vehicle Standard, "Battery Terminology," SAE Standard J1715/2, J1715/2_202108, August 2021.

[7.14].   SAE International Ground Vehicle Standard, "Electric-Drive Battery Pack System: Functional Guidelines," SAE Standard J2289, J2289_202108, August 2021.

[7.15].   SAE International Ground Vehicle Standard, "Industry Review of xEV Battery Size Standards," SAE Standard J3124, J3124_201806, June 2018.

[7.16].   SAE International Ground Vehicle Standard, "Best Practices for Storage of Lithium-Ion Batteries," SAE Standard J3235, March 2023, https://doi.org/10.4271/J3235_202303.

[7.17].   Bavishi, H., D'Arpino, M., Ramesh, P., Guezennec, Y. et al., "Comparative Analysis of Protection Systems for DC Power Distribution in Electrified Vehicles," SAE Technical Paper 2022-01-0135 (2022), doi:https://doi.org/10.4271/2022-01-0135.

[7.18].   Bisschop, R., Willstrand, O., Amon, F., and Rosengren, M., "Fire Safety of Lithium-Ion Batteries in Road Vehicles," RISE Research Institutes of Sweden, 2019, https://doi.org/10.13140/RG.2.2.18738.15049.

[7.19].   Chang, C., Gorin, C., Zhu, B., Beaucarne, G. et al., "Lithium-Ion Battery Thermal Event and Protection: A Review," *SAE Int. J. Elec. Veh.* 13, no. 3 (2024): 1-41, doi:https://doi.org/10.4271/14-13-03-0019.

[7.20].   SAE International Ground Vehicle Standard, "High Voltage Automotive Wiring Assembly Design," SAE Standard J1673, J1673_201203, March 2012.

[7.21].   Ford Motor Corporation, "2018 Focus Electric Battery Removal Guide," December 2017.

[7.22].   Kimoto, M., "Impact of Manual Service Disconnect in an Automotive Traction Battery System (RESS)," SAE Technical Paper 2017-01-1195, 2017, https://doi.org/10.4271/2017-01-1195.

[7.23].   SAE International Ground Vehicle Standard, "SAE Electric Vehicle and Plug In Hybrid Electric Vehicle Conductive Charge Coupler," SAE Standard J1772, J1772_201710, October 2017.

[7.24].   SAE International Ground Vehicle Standard, "NACS Electric Vehicle Coupler," SAE Standard J3400, J3400_202312, December 2023.

[7.25].   UL 2202, "DC Charging Equipment for Electric Vehicles," Edition 3, Underwriters Laboratories, December 2022.

[7.26].   SAE International Ground Vehicle Standard, "SAE Electric Vehicle Inductively Coupled Charging," SAE Standard J1773, J1773_201406, June 2014.

xEV Specific Risks

8.1.
Introduction

This chapter dissects the unique dimensions of fire risks inherent to xEVs, distinct from those associated with ICEVs previously covered in preceding chapters. Offering a dedicated exploration, it delves into the intricacies of mitigating, managing, and analyzing fire incidents specific to the domain of vehicles with electrified powertrains [8.1-8.5].

The journey unfolds by revisiting the pivotal role of high voltages that characterize xEVs, setting the stage for a comprehensive examination. Subsequently, the focus narrows to the challenges entwined with lithium batteries—the pulsating heart of current xEVs. This section unravels safety and optimal operating windows while delving into intricate phenomena such as

dendritic growth, internal short circuits, and thermal runaway, which are central to understanding most xEV-related fires.

As the narrative progresses, the analysis broadens to encompass the energy released during a fire, probing into its diverse facets. The chapter further navigates through the labyrinth of xEV particulars, addressing factors ranging from inherent complexity and manufacturing intricacies to challenges posed by recharging, collisions, flooding, submersion, toxic gases, and jet fires. Firefighting and extrication challenges are examined, shedding light on the distinctive difficulties encountered in emergency scenarios and some of the new technologies available for these situations. Lastly, the chapter touches on the concept of reignition and concludes by accentuating the critical aspects of toxic and hazard risks during the analysis of xEV fires.

8.2.
High Voltages

Elevated voltages in xEVs pose significant safety risks due to their presence in crucial components such as the high-voltage harness, high-voltage modules, traction battery, and supercapacitors. While safety devices are implemented to mitigate these risks, failures can occur, particularly after collisions or immersion. It is essential to recognize that dangerous electric energy may persist in specific xEV components for an extended period, potentially lasting months after the vehicle has been turned off. This extended presence of high voltages underscores the need for cautious handling and safety protocols to prevent potential hazards during maintenance, repair, emergencies, or analysis involving xEVs.

8.3.
Lithium Battery

8.3.1.
Safety and Optimal Operating Windows

Traction batteries and their control systems are meticulously designed to ensure the battery operates within an optimal range, aiming to extend its lifespan and enhance overall efficiency. This operational range is typically represented by a minimum and maximum operating temperature on one axis and the state of charge or voltage on the other, as illustrated in **Figure 8.1**. Additionally, there is a safety window surrounding this operational range. Any breach of the safety window, exceeding either the upper or lower limit for a specific cell, poses a substantial risk of permanent damage or catastrophic failures. Such breaches may result in internal short circuits or increased internal

temperature through various mechanisms. Consequently, the battery cell might experience failure, leading to the release of flammable gases, combustion, or even explosion [8.6-8.9].

Figure 8.1 Battery operating windows.

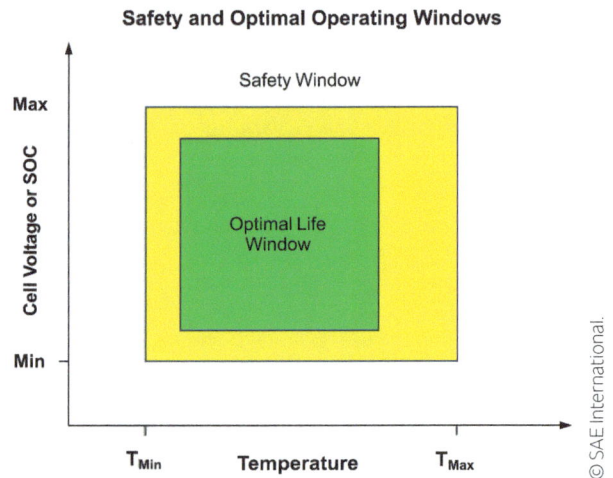

© SAE International.

Given that xEVs typically contain hundreds or even thousands of cells in their battery systems, a practical compromise is often made by recognizing the technical and economic challenges of monitoring and controlling each individual cell. The BMS selectively implements specific measurements only at a submodule level, which creates the possibility that a failure in a single cell may go unnoticed.

8.3.2.
Dendritic Growth

Dendritic growth refers to the development of slender, branching structures (dendrites) of lithium metal on the surface of the anode. This phenomenon is common in specific battery chemistries, particularly those using lithium metal or lithium-ion technology, and is exacerbated by high recharge currents and low electrolyte temperatures.

In **Figure 8.2**, the mechanism is graphically explained. A pristine lithium battery initially shows smooth anode and cathode surfaces, separated by a porous electrode embedded in a liquid electrolyte containing lithium salts and solvents. Lithium ions migrate through the electrolyte during charging, bonding as metallic lithium on the anode surface. Over multiple cycles, lithium atoms accumulate irregularly on the electrode surface, forming clusters that evolve into dendrites, extending toward the opposite electrode. If dendrites grow sufficiently, they can breach the porous electrode, reaching the battery anode and causing a metallic short circuit between the poles. This often leads to overheating, thermal runaway, fires, and explosions.

Researchers and engineers actively explore advanced battery technologies and materials to mitigate dendritic growth, considering different electrolyte formulations, electrode coatings, and innovative separator materials. Solid-state lithium batteries are also a potential solution, with intense development efforts. Addressing dendritic growth is crucial for ensuring lithium batteries' reliability, safety, and longevity in automotive applications, directly impacting electrified vehicles' performance and overall safety.

Figure 8.2 Dendritic growth along recharging cycles.

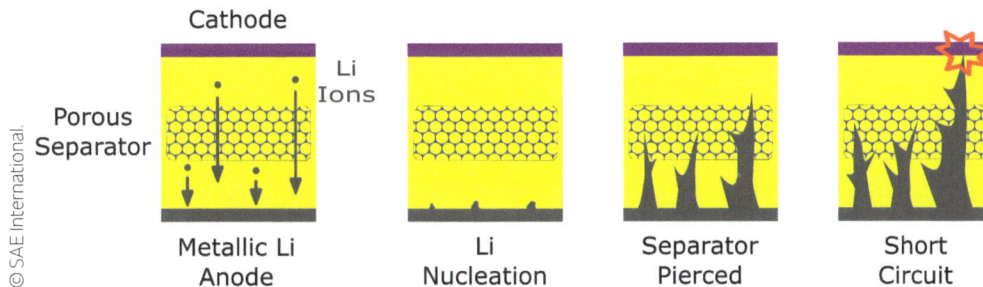

© SAE International.

8.3.3.
Thermal Runaway

Thermal runaway refers to an autonomous and uncontrolled increase in temperature within the battery cell, leading to catastrophic consequences. Exothermic degradation reactions occur as the electrolyte temperature surpasses a critical point, releasing heat. This further raises the temperature, triggering more degradation reactions and creating a positive feedback loop. Even a slight increase in electrolyte temperature significantly accelerates the speed of these reactions, resulting in the rapid generation of gases, vapors, heat, and pressure.

The polymeric separator may shrink in this process, allowing a short circuit within the cell's anode and cathode. While safety valves are designed to vent excessive pressure, releasing gases, vapors, and concomitant short circuits can initiate a fire or explosion. Surrounding cells become overheated and mechanically stressed, propagating a chain destruction mechanism. Some by-products of the degradation reactions act as oxidizers, sustaining flames even without fresh oxygen from the atmosphere. **Figure 8.3** illustrates these complex interactions, emphasizing the critical importance of robust safety mechanisms and materials to prevent or mitigate these dangerous events.

Figure 8.3 Thermal runaway.

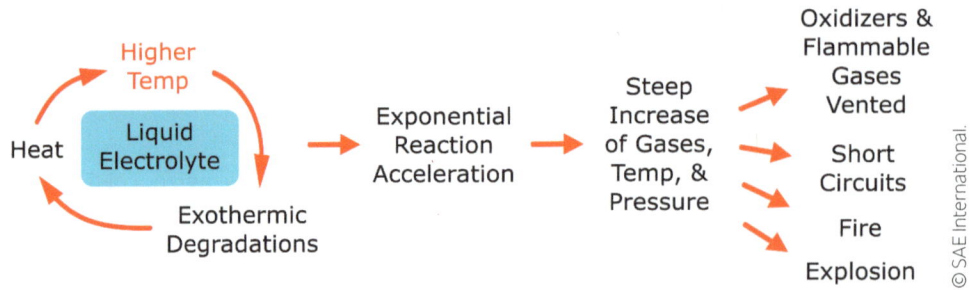

According to studies, including the work by Chang et al. [8.7], a gradual sequence of thermal events leads to the ultimate occurrence of thermal runaway. This progression is intricately elucidated in the diagram presented in **Figure 8.4**.

While thermal runaway is frequently instigated by dendritic growth and the ensuing internal short circuit within an individual cell of the battery system, various factors can contribute to this undesirable outcome. Causes may stem from manufacturing issues, such as contaminants,

Figure 8.4 Stages of LIB thermal events.

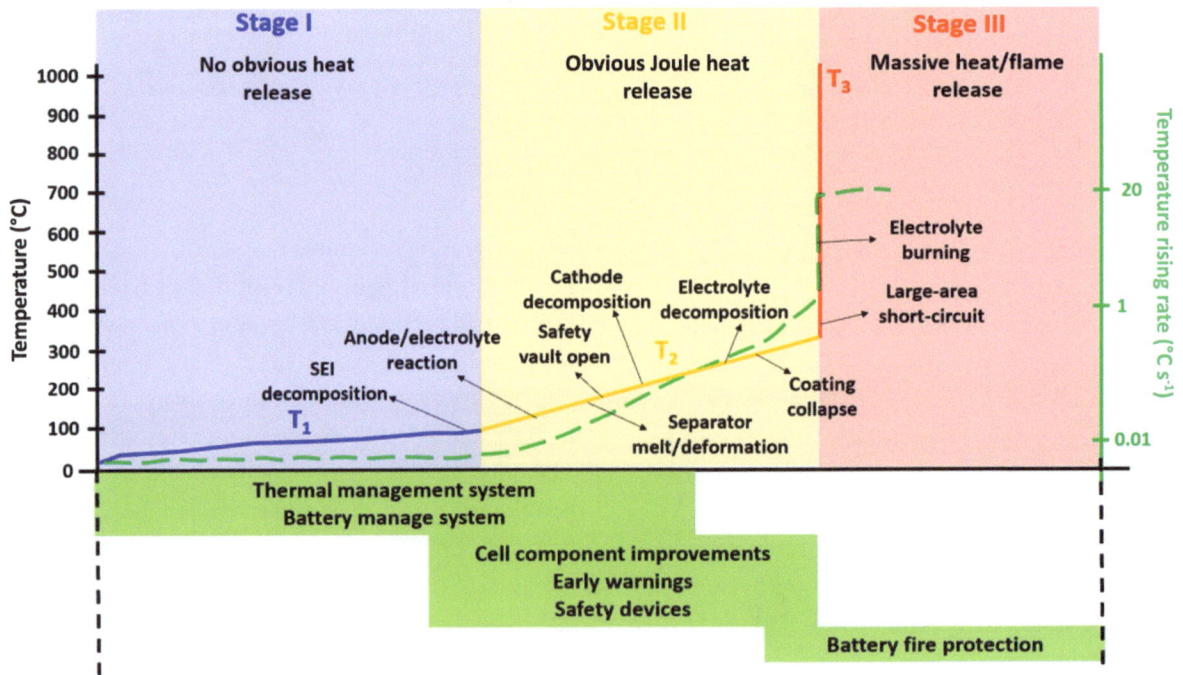

electrode burrs, and misalignments, as well as environmental factors like exposure to extreme temperatures. Additionally, electrical abuse, which theoretically should be prevented by the

BMS, and mechanical damage can also play a role. Specific incidents result from a sequence of events, as illustrated in **Figure 8.5**.

Figure 8.5 Common causes of LIB thermal runaway.

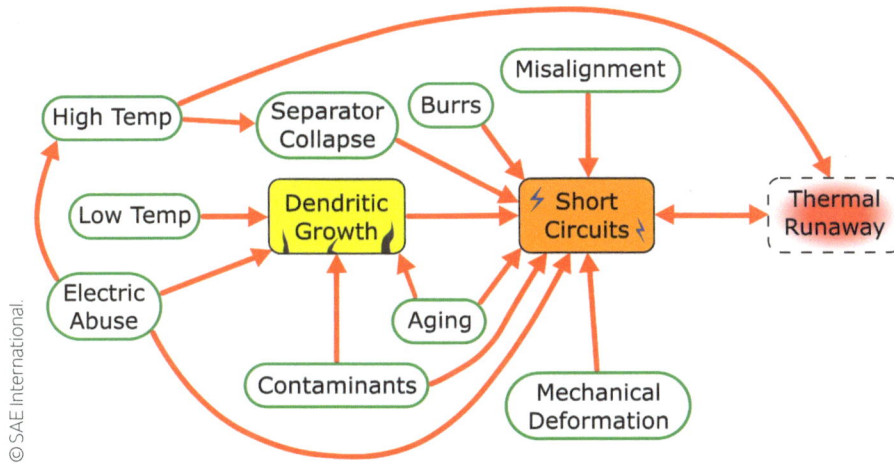

© SAE International.

8.3.4.
Energy Released in a Fire

Contrary to the intuitive notion that the thermal energy released by a lithium battery can be explained mainly by the electric energy it has stored as an electric charge, the reality is more complex. Lithium batteries exhibit three distinct mechanisms that ultimately translate to thermal energy: (1) electric energy, (2) exothermal reactions, and (3) burning of its materials and by-products [8.10]. This is further detailed in the following sequence.

8.3.4.1.
Heat from Electric Energy

In the worst case, all the electric energy stored in the battery contributes to elevating the temperature of surrounding materials through the Joule effect or short circuits and arcing. Since some

electrolyte by-products conduct electricity, the leakage paths within a faulty cell will be heated by the square of the crossing current times the equivalent resistance, producing a relatively modest temperature increase. On the other hand, short circuits and the associated arcing can exceed several thousand degrees Fahrenheit and Celsius in the arcing path, quickly igniting combustions.

8.3.4.2.
Heat from Exothermic Reactions and Adjacent Cells

In addition to the electric contribution to heat generation examined in the previous item, an LIB fire involves heat generated internally within the cell due to complex exothermic reactions of the overheated electrolyte and eventually from reactions with the substances from the battery electrodes. Furthermore, a damaged cell acts as a thermal energy source in the battery pack of typical electrified vehicles, overheating surrounding cells.

8.3.4.3.
Heat from Burning Fuels

This energy segment arises from the combustion of flammable materials, either in combination with oxygen from the atmosphere or with oxidizers from the LIB.

As Sun et al. elucidate [8.4] and as depicted in **Figure 8.6**, the thermochemical energy released during a battery fire, encompassing both the heat generated by thermal runaway inside the battery and the sustenance of flames by flammable gases expelled from the battery, far exceeds the electrical energy stored in the battery. The LIB fire has the potential to release 5–10 times more energy than the stored electrical energy, contingent upon its SOC.

Figure 8.6 Energy release in LIB fire.

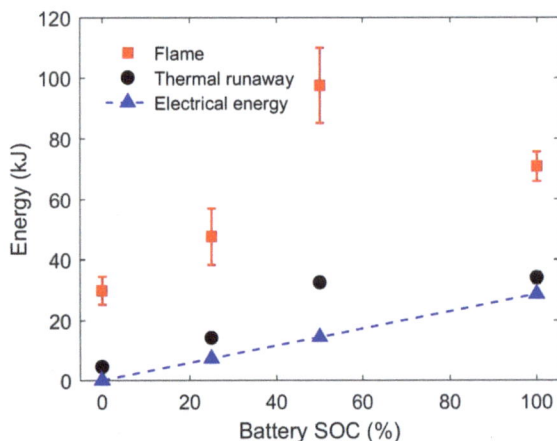

Used with permission of Kluwer Academic Publishers (Boston), from A Review of Battery Fires in Electric Vehicles, Sun, Peiyi & Bisschop, Roeland & Niu, Huichang & Huang, Xinyan, 1-50, 2020; permission conveyed through Copyright Clearance Center, Inc.

8.3.4.4.
Fuels

The liquid electrolytes of LIBs typically contain flammable substances such as organic solvents. Additionally, some of the by-products resulting from degradation reactions are also flammable. The porous separator is usually a combustible polymer. It is important to recall that metallic lithium, in the presence of oxygen or other oxidizers, is also flammable, as are some other substances used in the anode and cathode of LIB. Plastic enclosures (of the cell and of the battery module) can also serve as fuels for an already established fire.

8.3.4.5.
Oxidizers

Unfortunately, several substances used in the anode are oxidizers or can release oxygen, fluor, and free radicals during a thermal runaway. Additionally, some by-products of the degradation of the electrolyte can act as oxidizers. This explains why the LIB fire does not need oxygen from the atmosphere to continue once ignited. It also clarifies why traditional firefighting substances like CO_2, water, and others are less effective compared to their use in extinguishing an ICEV fire.

8.3.5.
LIB Fire Triangle

The fire triangle model can be specifically applied to LIBs, as depicted in **Figure 8.7**. In this context, the fuel side includes the electrolyte, anode, separator, and plastic enclosure. Heat sources can be either internal or external, often involving short circuits. Oxidizers are not limited to atmospheric oxygen (O_2); they also include various electrolyte by-products such as lithium manganese oxide (LMO), lithium cobalt oxide (LCO), nickel manganese cobalt (NMC), nickel cobalt aluminum (NCA), and lithium iron phosphate (LFP).

Figure 8.7 LIB fire triangle.

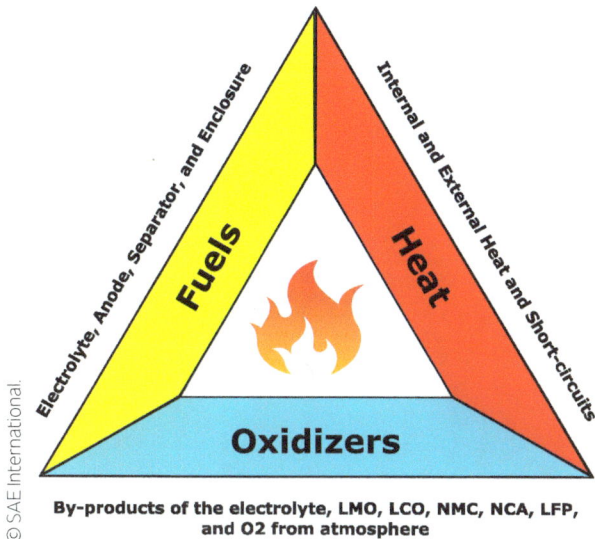

By-products of the electrolyte, LMO, LCO, NMC, NCA, LFP, and O2 from atmosphere

8.4.
Complexity

From the currently produced xEV topologies, full HEVs stand out for their unparalleled complexity. This complexity arises from the intricate integration of powertrain systems found in ICEVs, encompassing components like the engine, fuel tank, and engine control module. Additionally, full HEVs incorporate elements from BEVs, including the traction battery, the electric motor, and dedicated control modules, adding another layer of technical intricacy involving various management systems and modules.

Considering the reliability aspect, it is well-established that the greater the complexity of a real-life system, the higher the likelihood of failures unless effective countermeasures are implemented. Notably, a fraction of these failures

potentially escalate to car fires. Recent data on car fires, as detailed in Chapter 1 of this book, substantiate this viewpoint, revealing higher fire rates for HEVs in comparison to both ICEVs and BEVs. This underscores the critical need to address reliability concerns and proactively implement measures to mitigate failures and associated safety risks in the more complex variants of HEVs.

8.5.
Manufacturing and Service

The manufacturing and servicing of xEVs pose distinctive risks that warrant careful consideration to safeguard the well-being of operators and technicians. These risks encompass exposure to high voltage, potential chemical hazards associated with battery handling, and the fire and thermal risks linked to LIBs. Additionally, the specialized nature of high-voltage systems necessitates using specific tools and equipment.

To effectively mitigate these risks, it is imperative to implement comprehensive safety measures. This includes the adoption of specialized personal protective equipment (PPE), as illustrated in **Figure 8.8**. Unlike traditional automotive PPE, certain job positions may require high-voltage gloves, face shields, and insulated boots. Stringent safety protocols must be established, and specialized training programs for operators and technicians should be provided. Furthermore, facilities should be designed to accommodate the unique characteristics of electrified vehicles [8.11].

Figure 8.8 PPE examples.

8.6.
Recharging

Recharging stations employing cables and plugs incorporate safety features. Electric energy flow is allowed only when the connector is fully fixed and a communication protocol verifies the vehicle's energy-related details. The plug is mechanically designed for user protection and prevents particle and water ingress when connected. Similarly, induction rechargers activate the transmitter coil only when a compatible vehicle is aligned correctly.

Despite these built-in safety features, recent fires at recharging stations and in home garages indicate the need for improved automatic fire detection, countermeasures, and response times. This is crucial to prevent fires initiated during vehicle recharge from escalating and causing damage to nearby facilities and vehicles, sometimes in a domino effect, as implied in **Figure 8.9**.

Figure 8.9 Domino effect in a charging station.

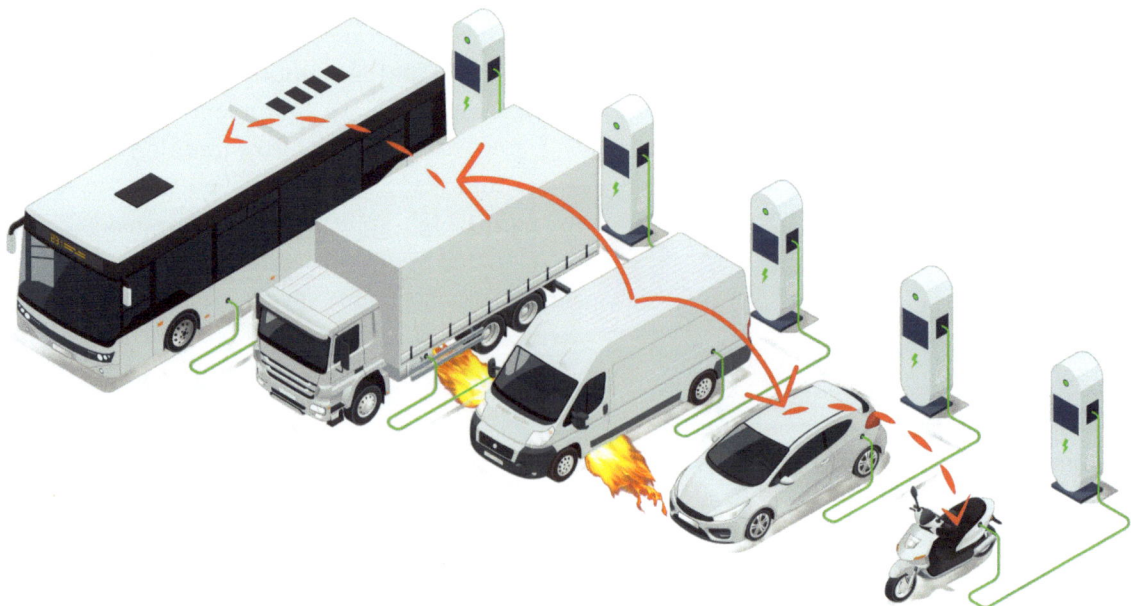

A similar incident in May 2020, captured by a surveillance camera in China, resulted in fire damage to at least three vehicles parked side by side at a charging station [8.12]. A thermal runaway in the first vehicle, a minivan, spread to a small truck in the adjacent bay, resulting in a battery fire that damaged a nearby sedan. In July 2021, GM recommended owners of 2017–2019 Chevrolet Bolts to park them outdoors and refrain from overnight charging or reaching a 100% charge due to two electric cars catching fire after recall repairs. A second recall was issued, including a software change limiting the battery charge to 80% [8.13].

8.7.
Collisions

With the increasing daily use of xEVs and ongoing technological advancements, new safety incidents arise, necessitating continuous improvements in design and regulations. In the US, collision regulations initially created for ICEVs were found inadequate for addressing certain risks linked to large traction batteries. This led to the introduction of FMVSS 305 in September 2017, initially focusing on electric-powered vehicles, and later expanded in August 2019 to include hydrogen fuel cell vehicles and 48 V mild hybrid technologies [8.14].

In 2013, two unique collisions involving Tesla Model S vehicles resulted in underbody damage, leading to car fires. One of these incidents involved the battery pack being penetrated by metallic debris on the road, causing battery damage and a subsequent fire. Subsequently, in 2014, Tesla enhanced its Model S vehicles by adding a titanium underbody shield and aluminum deflector plates to prevent damage and fires resulting from underbody

collisions with unusual debris. These enhancements were retrofitted to all cars manufactured after a specific date, and Tesla service also offered retrofits to existing cars upon request or as part of regular service. According to Tesla's website, extensive vehicle-level tests demonstrated that these enhancements effectively prevented any damage that could lead to a fire or breach the existing quarter-inch ballistic-grade aluminum armor plate protecting the battery pack [8.15].

Collisions involving xEVs require distinct protocols compared to those for ICEVs, considering the specific risks of electric shocks, electrocution, exposure to toxic substances, and the potential for first responders to cause unintentional fires. The NFPA's "Alternative Fuel Vehicles Safety Training Program Emergency Field Guide" is a valuable resource for firefighters and first responders, providing essential information to minimize the risks of handling xEV collisions. Readers are also encouraged to get familiar with the guides issued by IAFC, ISO, NHTSA, NTSB, and the US Fire Administration [8.16-8.28].

8.8.
Flooding and Submersion

Instances of xEVs igniting after exposure to salty water during hurricanes Sandy (2012) and Ian (2022) highlight the imperative to enhance vehicles and protocols for addressing flooding. Residual salt in the battery and high-voltage components can create conductive "bridges," leading to short circuits and self-heating and potentially resulting in fires [8.29]. The ignition time frame for a damaged battery can vary widely, ranging from days to weeks. The NHTSA recommends promptly identifying flooded xEVs and relocating them at least 50 ft away from structures, other vehicles, or combustibles.

Furthermore, incidents where xEVs submerge in water bodies (e.g., lakes, rivers, and seawater) can establish hazardous current paths from damaged high-voltage components, impacting occupants or first responders. Notably, accidental immersion of xEVs, even without a collision, has triggered some fires, as demonstrated by incidents where vehicles immersed while towing a boat ramp down [8.30, 8.31]. **Figure 8.10** depicts a flooded BEV experiencing electrical or battery damage. This damage generates combustible gases, which are visible burning above the water level.

Figure 8.10 Fire of an immersed BEV.

Fresh_Studio/Shutterstock.com.

Guidelines referred to in the preceding discussion alert involved personnel to these dangers and provide instructions on recognizing signs indicating it is unsafe to enter the water near the vehicle. Importantly, it is worth noting that metallic lithium, present in varying quantities inside some LIBs, reacts violently when exposed to water, potentially triggering a fire from the generated gases.

8.9.
Toxic Gases and Jet Fires

Fires involving the traction batteries or supercapacitors of xEVs introduce the potential emission of toxic gases and the formation of jet fires. In a fire, these components can release harmful substances such as carbon monoxide, hydrogen fluoride, hydrogen cyanide, and hydrogen chloride, posing inhalation risks for first responders and those in proximity. Moreover, as shown in **Figure 8.11**, jet fires may extend several feet (a few meters) horizontally during combustion. This phenomenon intensifies the fire scene and heightens the risk of flames reaching nearby vehicles and structures, rapidly escalating the incident's impact.

Figure 8.11 Jet fire from LIB.

Whitevector/Shutterstock.com.

8.10.
Firefighting Challenges

While studies suggest that the energy released in xEV fires is comparable to ICEV fires, addressing xEV incidents presents more formidable challenges. In contrast to conventional ICEV fires, where tactics such as water, foam, or blankets prove effective in mitigating contact with oxygen, xEV fires fueled by LIBs in thermal runaway generate their oxidizers internally, rendering conventional methods ineffective. Furthermore, in contrast to ICEVs, most xEV incidents manifest jet fires, and occasionally, the

flammable gases released during an LIB thermal runaway can lead to an explosion before visible flames emerge.

Complicating matters, LIB modules' metallic and plastic casings hinder temperature reduction through water spraying, as water may not directly reach all hot battery cells. Consequently, firefighting efforts often necessitate a larger volume of water and lead to prolonged activities. While ICEV fires are usually extinguished in less than 45 min, xEV fires might require several hours. The International Association of Fire Chiefs (IAFC) currently recommends securing a continuous and sustainable water supply ranging from 3000 to 8000 gal (14,000–36,000 L).

Faced with these difficulties, some proponents suggest allowing the vehicle to burn until all combustible materials, including the LIB cells, are entirely consumed. This approach may present a more straightforward solution, particularly in open spaces and when the vehicle is isolated from other vehicles or structures.

Innovative methods are being introduced with some success in response to the xEV fire challenges [8.32-8.34]. One such approach involves deploying an "inverted shower" nozzle. The example seen in **Figure 8.12** is a device made by Turtle Fire Systems™, which can be easily positioned under the vehicle to direct water toward the battery from beneath. The flow rate can be adjusted from 75 to more than 500 GPM (280 to 1900+ L/min), by adjusting the pressure at the fire pump, therefore adapting to water availability and the different stages of a thermal runaway.

Figure 8.12 Turtle Fire System™.

Courtsey of Turtle Fire Systems.

Another strategy involves tilting the vehicle to enable a more direct water application over the extensive surface of the battery module beneath, using conventional nozzles (see **Figure 8.13**). Despite the need for large jacks and substantial water amounts, this method represents an alternative.

Figure 8.13 BEV tilted for water access.

Courtesy of Metro Fire of Sacramento.

An additional technique involves perforating the battery enclosure through the vehicle floor using a hollow metallic lance and a sledgehammer, as illustrated in **Figure 8.14**. This allows continuous water injection, requiring a smaller volume and less time to extinguish the LIB fire and cool the battery cells. However, this method raises concerns about creating new short circuits, exposing firefighters to high voltage, and the need for their proximity until the lance penetrates the required spot. Additionally, the internal arrangement of the battery submodules might prevent the external water from effectively cooling all the necessary cells.

A similar concept in **Figure 8.15** features an apparatus designed to pierce the battery system from beneath the vehicle and inject a moderate flow of water. Developed by Rosenbauer [8.32], the piercing stinger can be activated safely from a distance of 25 ft (7.5 m) and may require as little as 500 gal (1900 L) of water to extinguish a vehicle battery fire. Despite these advancements, challenges remain, particularly in addressing the diverse designs of battery systems. The internal construction of some battery systems may restrict the flow of the injected water, preventing effective cooling of all the cells.

Some companies promote and offer a range of products designed to, at least, contain, retard, or isolate xEV fires while awaiting the arrival of firefighters or better-equipped response teams. Their set of products may include a fire blanket, a water mist lance (utilizing water mist, which, while typically unable to extinguish an LIB fire, effectively reduces temperature and restricts oxygen access more efficiently than a standard water spray), aerosol deployers (to disrupt the chain reaction of flames external to the LIB), a water injection lance, and an insulated sledge-hammer. These tools are intended to provide an initial response to mitigate the impact of an xEV fire and enhance safety measures until professional firefighting resources arrive at the scene.

Figure 8.14 Water injection through lance. (a) Hammer and lance (b) Perforation of the vehicle floor (c) Lance in place.

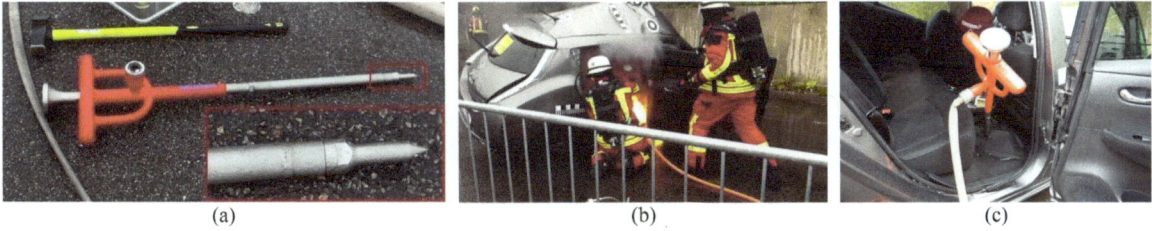

(a)   (b)   (c)

Reprinted from SAE Technical Paper 2021-01-0847. © SAE International.

Figure 8.15 Water injection system. (a) Device being displaced below the vehicle floor (b) Stinger elevated for illustration.

(a)

© Rosenbauer International AG.

(b)

8.11.
Extrication Challenges

The extrication of victims involved in xEV collisions presents a significant concern, demanding specialized knowledge and procedures. Firefighters now train to safely disconnect high-voltage electric systems before engaging with the vehicle. Additionally, they must identify which parts of the vehicle structure can be safely cut and stretched using familiar tools employed in ICEV accidents, ensuring that these actions do not lead to short circuits or expose victims to the risk of electric shock [8.35].

Addressing these challenges, the SAE J3108 Ground Vehicle Standard has been developed to standardize labels, badges, and other indicative elements in xEVs [8.26]. This standardization aims to streamline and expedite the efforts of firefighters and first responders by providing clear guidance on how to approach xEV extrication scenarios.

Many xEV manufacturers collaborate with the NFPA to further support emergency responders, offering Emergency Field Guide materials [8.21]. These guides provide essential information for responding to emergencies and offer simplified diagrams that readily identify crucial components such as the MSD plug, traction battery, airbags, high-voltage harnesses, and other safety-critical components.

8.12.
Reignition

Reignition and stranded energy pose critical challenges in xEV fires, with secondary thermal events potentially occurring hours, days, or weeks after the initial LIB fire has been extinguished. Notable instances of reignition have been documented across various vehicles, such as a Tesla Model S crash in Florida, US, in May 2018, where the car, after being engulfed in flames, experienced reignition both at the accident site and after arrival at the tow yard.

The primary factor contributing to these reignition scenarios is stranded energy, which refers to electric energy remaining in the battery or supercapacitors. Additionally, degraded materials and new short circuits further support the reignition process. This complexity raises concerns for postcrash handlers, who often lack the tools or training to manage such events safely. Recognizing and addressing these factors are imperative for developing effective safety measures for postcrash fires involving xEVs.

Current recommendations from the IAFC, NFPA, and NHTSA advocate for the safe storage of damaged LIBs, wrecks of xEVs with this battery type, or xEVs exposed to flooding in a designated zone at least 50 ft (15 m) away from any structure and other combustible materials, as illustrated in **Figure 8.16** [8.36].

In the Netherlands, firefighters occasionally employ tow trucks to convey sizable shipping containers to the location of xEV accidents involving compromised batteries. **Figure 8.17** shows one example of a container for that use. These containers are filled with water, and a small crane raises the vehicle, lowering it into the bath for safe immersion. Consequently, the battery is flooded, cooled, and discharged, preventing potential reignition. This approach is becoming increasingly prevalent across Europe.

Figure 8.16 Safety perimeter around xEV ruins.

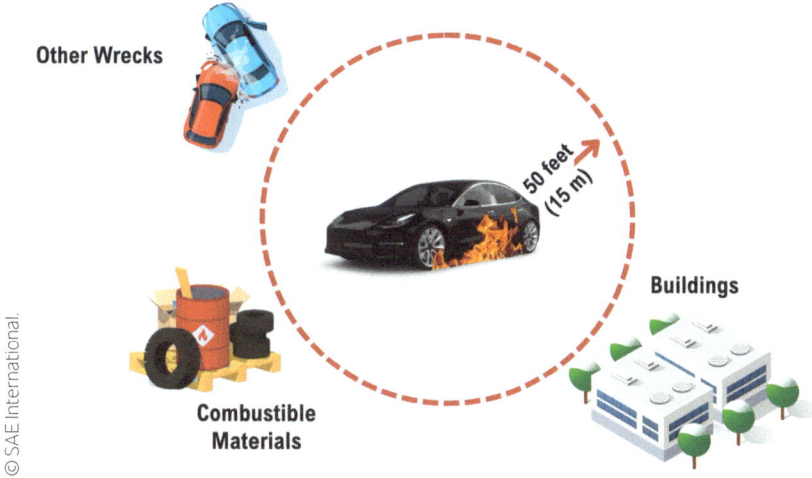

© SAE International.

Figure 8.17 Container for soaking a fired xEV in water.

TLF/Shutterstock.com.

8.13.
(Toxic and Hazard Risks during) Fire Analysis

Even if an xEV fire analysis by a forensic engineer or a similar professional is conducted several days after the incident, the consideration of stranded energy remains imperative. Consequently, the risks associated with high voltages and potential new thermal runaways should always be considered. It is crucial to note the likely presence of toxic or harmful by-products of the fire during xEV analysis, emphasizing the importance of avoiding inhalation and dermal contact with these substances.

While ICEV fires also require proper PPE and caution from vehicle examiners due to the diverse materials used in modern cars, the analysis of xEV fires amplifies these risks. This heightened concern arises from the involvement of lithium batteries and, in some cases, supercapacitors. In essence, as emphasized by Mr. Ofodike Ezekoye in a recent NFPA Webinar [8.17], "caution should be exercised to limit inhalation of suspended products and dermal contact with debris."

References

[8.1].   Blikeng, S. and Agerup, S., "Fire in Electric Cars," SAE Technical Paper 2015-01-1382 (2015), doi:https://doi.org/10.4271/2015-01-1382.

[8.2].   EV Fire Safe, "Risks with EV Battery Fire," EV Fire Safe Is Supported by the Australian Government, Department of Defence, accessed January 2024, https://www.evfiresafe.com/risks-ev-fires.

[8.3].   National Highway Traffic Safety Administration (NHTSA), "Interim Guidance for Electric and Hybrid-Electric Vehicles Equipped with High Voltage Batteries," January 2012, accessed January 2024, https://www.nhtsa.gov/sites/nhtsa.gov/files/interimguide_electrichybridvehicles_012012_v3.pdf.

[8.4].   Sun, P., Bisschop, R., Niu, H., and Huang, X., "A Review of Battery Fires in Electric Vehicles," *Fire Technology* 56 (2020): 1361-1410, doi:https://doi.org/10.1007/s10694-019-00944-3.

[8.5].   Takahashi, M., Takeuchi, M., Maeda, K., and Nakagawa, S., "Comparison of Fires in Lithium-Ion Battery Vehicles and Gasoline Vehicles," *SAE Int. J. Passeng. Cars - Electron. Electr. Syst.* 7, no. 1 (2014): 213-220, doi:https://doi.org/10.4271/2014-01-0428.

[8.6].   Bisschop, R., Willstrand, O., Amon, F., and Rosengren, M., "Fire Safety of Lithium-Ion Batteries in Road Vehicles," RISE Research Institutes of Sweden, 2019, https://doi.org/10.13140/RG.2.2.18738.15049.

[8.7].   Chang, C., Gorin, C., Zhu, B., Beaucarne, G. et al., "Lithium-Ion Battery Thermal Event and Protection: A Review," *SAE Int. J. Elect. Veh.* 13, no. 3 (2024): 1-41, doi:https://doi.org/10.4271/14-13-03-0019.

[8.8].   Qi, C., Liu, Z., Lin, C., and Hu, Y., "Review of Gas Generation Behavior during Thermal Runaway of Lithium-Ion Batteries," *SAE Int. J. Elec. Veh.* 13, no. 3 (2024): 1-15, doi:https://doi.org/10.4271/14-13-03-0021.

[8.9].   SAE International Ground Vehicle Standard, "Best Practices for Storage of Lithium-Ion Batteries SAE J3235," SAE Standard J3235, March 2023, https://doi.org/10.4271/J3235_202303.

[8.10]. Roman, J., "Stranded Energy," National Fire Protection Association Journal, January 1, 2020, accessed January 2024, https://www.nfpa.org/news-blogs-and-articles/nfpa-journal/2020/01/01/ev-stranded-energy.

[8.11]. SAE International Ground Vehicle Standard, "Hybrid and Electric Vehicle Safety Systems Information Report," SAE Standard J2990/2, J2990/2_202011, November 2020.

[8.12]. CCTV Images (Related to 4 EVs Damaged on a Recharging Station in China, on May 5th, 2020), "Direct Hit: Electric Car Explodes While Charging, Then Burns Up," https://www.youtube.com/watch?v=IIHv32QXVXE; Duplicate Footage Appeared on Several Sites, Such As, accessed January 2024, https://www.newsflare.com/video/355696/four-vehicles-burnt-after-fire-erupts-from-electric-car-charging-at-station-in-china#, https://twitter.com/PDChina/status/1259211538981810176.

[8.13]. CBS News, "General Motors Tells Chevy Bolt Owners to Park Outside Because Batteries Could Catch Fire," accessed January 2024, https://www.cbsnews.com/news/chevy-bolt-ev-recall-battery-warning-fire-charging-park-outdoors/.

[8.14]. National Highway Traffic Safety Administration (NHTSA), "Electric-Powered Vehicles: Electrolyte Spillage and Electrical Shock Protection," FMVSS 305, Code of Federal Regulations, Federal Motor Vehicle Safety Standards, 49 CFR 571.305 Standard No. 305, August 2019.

[8.15]. Tesla Blog, "Tesla Adds Titanium Underbody Shield and Aluminum Deflector Plates to Model S," accessed January 2024, https://www.tesla.com/blog/tesla-adds-titanium-underbody-shield-and-aluminum-deflector-plates-model-s.

[8.16]. Egelhaaf, M., Ruecker, P., and Heyne, T., "Firefighting of Li-Ion Traction Batteries - An Update," *SAE Int. J. Adv. & Curr. Prac. in Mobility* 3, no. 6 (2021): 3085-3092, doi:https://doi.org/10.4271/2021-01-0847.

[8.17]. Ezekoye, O., "Firefighter Safety on Firegrounds Involving Lithium-Ion Batteries," NFPA Webinar, NFPA Fire Protection Research Foundation, November 20, 2023, https://www.nfpa.org/videos/firefighter-safety-on-firegrounds-involving-lithium-ion-batteries.

[8.18]. Ghiji, M., Novozhilov, V., Moinuddin, K., Joseph, P. et al., "A Review of Lithium-Ion Battery Fire Suppression," *Energies* 13, no. 19 (2020): 5117, doi:https://doi.org/10.3390/en13195117.

[8.19]. International Association of Fire Chiefs (IAFC) Bulletin, "Fire Department Response to Electrical Vehicle Fires," October 15, 2021, accessed January 2024, https://www.iafc.org/docs/default-source/1haz/respondingtoelectricalvehiclefires.pdf?sfvrsn=9421650c_6.

[8.20]. International Standards Organization, "Road Vehicles—Information for First and Second Responders—Part 1: Rescue Sheet for Passenger Cars and Light Commercial Vehicles," ISO Standard 17840-1:2022(en), 2022.

[8.21]. National Fire Protection Association, "NFPA's Alternative Fuel Vehicles Safety Training Program Emergency Field Guide," 2018, ISBN:978-1455912742.

[8.22]. National Fire Protection Association (NFPA), "National Fire Protection Association First Respond Video," accessed January 2024, https://www.youtube.com/watch?v=nN-RUJ0a4K4.

[8.23]. National Highway Traffic Safety Administration (NHTSA), "Interim Guidance for Electric and Hybrid-Electric Vehicles Equipped with High-Voltage Batteries—(Law Enforcement/Emergency Medical Services/Fire Department)," accessed January 2024, https://www.nhtsa.gov/sites/nhtsa.gov/files/811575-interimguidehev-hv-batt_lawenforce-ems-firedept-v2.pdf.

[8.24]. National Transportation Safety Board (NTSB), "Safety Risks to Emergency Responders from Lithium-Ion Battery Fires in Electric Vehicles," accessed January 2024, https://www.ntsb.gov/safety/safety-studies/Documents/SR2001.pdf.

[8.25]. SAE International Ground Vehicle Standard, "Hybrid and EV First and Second Responder Recommended Practice," SAE Standard J2990, J2990_201907, July 2019.

[8.26]. SAE International Ground Vehicle Standard, "xEV Labels to Assist First and Second Responders, and Others," SAE Standard J3108, J3108_201703, March 2017.

[8.27]. Underwriters Laboratories (UL) and HowItBroke, "Electric Vehicle Firefighting Reminders," accessed January 2024, https://ul.org/sites/default/files/2021-04/EV_FF_Handout20210422.pdf.

[8.28]. US Fire Administration (USFA), "Emergency Response Guides for Electric Vehicles and Lithium-Ion Batteries," accessed January 2024, https://www.usfa.fema.gov/blog/ig-062322.html#Downloadable%20Ergs.

[8.29]. US Fire Administration (USFA) Blog, "Responding to Electric Vehicle Fires Caused by Salt Water Flooding," accessed January 2024, https://www.usfa.fema.gov/blog/ig-102022.html.

[8.30]. NBC Los Angeles, "Tesla Crashes into Lake Forest Home, Igniting Fire," https://www.nbclosangeles.com/news/tesla-lake-forest-home-fire/22740/, August 26, 2017, Unable to Access the Complete Video, Similar Youtube Video, accessed January 2024, https://www.youtube.com/watch?v=3eFM9JJMH_0.

[8.31]. The Sun US, "Flamin' Fury—Horrifying Moment Tesla Malfunctions and Plunges into Lake... Before Bursting into Flames Underwater," accessed January 2024, https://www.the-sun.com/motors/9603410/tesla-model-x-malfunctions-sinks-water-fire/.

[8.32]. Rosenbauer, "Battery Extinguishing System Technology," Rosenbauer, accessed January 2024, https://rosenbaueramerica.com/fire-trucks/rosenbauer-equipment/.

[8.33]. Central and West Brabant Fire Brigade, "Feu de voiture brand van een auto," accessed January 2024, https://oudpompierke.be/actualite/feu-de-voiture-brand-van-een-auto.

[8.34]. Fire Isolator, "Fighting and Controlling EV Fires," The Netherlands, accessed January 2024, https://fireisolator.com/.

[8.35]. Fire Engineering, "Fire Engineering Video," accessed January 2024, https://www.fireengineering.com/technical-rescue/extrication-zone/video-high-voltage-vehicle-firefighting/#gref.

[8.36]. National Highway Traffic Safety Administration (NHTSA), "Interim Guidance for Electric and Hybrid-Electric Vehicles Equipped with High-Voltage Batteries—(Towing and Recovery Operators and Vehicle Storage Facilities)," accessed January 2024, https://www.nhtsa.gov/sites/nhtsa.gov/files/811576-interimguidehev-hv-batt_towing-recovery-storage-v2.pdf.

Chapter 09

Fire Spread and Dynamics

9.1.
Introduction

This chapter builds on a foundational understanding of fire dynamics, particularly in the context of more common structural fires often found in residential settings. It proceeds to illuminate the distinctive attributes of automotive environments, paving the way for detailed exploration of specific scenarios, including fires that initiate in the engine compartment (a predominant occurrence in vehicle fire statistics), those originating within the cabin, and those sparked within the traction batteries of xEVs. The chapter also delves into critical considerations, such as instances where vehicle components may exhibit explosive behavior. Lastly, it introduces the intriguing question of causality in the context of the chemical versus electrical fire root cause, akin to the classic "chicken and egg" paradox. The effects of wind, fire evolution phases, and typical automotive fire patterns will be addressed throughout.

9.2.
General Case/Structural Fires

Let us consider an example of fire evolution in a room. The spread of fire is relatively intuitive, at least in its initial stages. Assuming a red-hot cigarette falls on a bed, as exemplified in **Figure 9.1**.

Figure 9.1 Lit cigarette close to bed linen.

yamasan0708/Shutterstock.com.

Figure 9.2 Flames from mattress combustion.

Vlad Kagoshima/Shutterstock.com.

Given adequate conditions, its ember will ignite the linen, and the situation might evolve to a more noticeable combustion, with flames and a more significant amount of smoke. From this point, the flame boundaries will progress radially in an expanding horizontal circle of destruction. At the same time, the combustion front will penetrate the mattress vertically, damaging cushion materials and foams. The flames will increase in height, as seen in **Figure 9.2**, eventually reaching curtains and nearby furniture. Some incandescent pieces may fall and ignite a carpet, further spreading the fire horizontally in the room.

The hot air and flames will move upward, reaching the ceiling, and then radially, eventually getting through to adjacent chambers. Wall paint, doors, and frames will heat up and ignite. If the room is in a multistory house (or a building), provided there is no sprinkler system or containment actions, the flames will likely continue to increase. At some point, they will shatter window glass (because of thermal shocks or thermal gradients) and move upward, spreading the fire to the floor above, as shown in

Figure 9.3. It may take a few hours for the fire to consume all the combustible materials of a typical house if firefighters do not act.

The well-known conduction, irradiation, and convection mechanisms can explain the heat spread. The temperature of the linen and air immediately surrounding the combustion front increases by conduction. Due to their high temperatures, the flames radiate heat to surrounding surfaces, also elevating their temperatures. The hot air (now mixed with other gases, vapors, and smoke) expands and moves upward (convection), transporting flames and creating a flow of fresh air (with oxidizing O_2 gas) that will reach unburnt combustibles.

Figure 9.3 Fire ascending.

sashkO/Shutterstock.com.

Also, some substances not burning at ambient temperature will contribute to the fire once heated enough (above the flash point). Other materials will decompose (pyrolyze) and provide combustible by-products. Exothermal reactions will bring more heat to the scene, allowing the fire to evolve until all combustibles or oxidizers are consumed [9.1].

9.3.
Phases and Special Events

In understanding automotive fires, it is helpful to recognize that they do not simply burn indiscriminately but follow a structured sequence of phases, each marked by unique behaviors and conditions. This systematic progression through phases is a fundamental framework to comprehend and effectively control fires. However, it is equally important to acknowledge that this progression is only sometimes linear, as special events can disrupt the otherwise

continuous journey of fire development. This section will delve into the typical phases of automotive fire development and explore the special events that can interject this seemingly predictable course.

9.3.1.
Fire Development Phases

Fires typically advance through various phases, each marked by distinct characteristics and conditions. These phases are valuable tools for aiding firefighters and safety experts in comprehending and effectively addressing fires. The following sections discuss the typical fire phases.

9.3.1.1.
Ignition Phase

This is the initial stage of a fire. It begins when a source of heat or ignition contacts a combustible material. In this phase, the fire is small and localized, and it usually requires adequate heat, oxygen, and fuel to continue and progress to the next stage.

9.3.1.2.
Growth Phase

Once ignited, the fire proliferates if it finds enough fuel and oxygen. The development of flames characterizes this phase, with the release of heat and smoke, and an increase in the size and intensity of the fire. The heat of the fire warms up nearby fuel sources, making them more susceptible to ignition.

9.3.1.3.
Fully Developed Phase

In this stage, the fire reaches its maximum intensity and size. The flames are at their most vigorous, and the heat release is at its peak. The fire spreads rapidly through available fuel, consuming it as it goes.

9.3.1.4.
Decay or Extinction Phase

As the fire consumes most of the available fuel or encounters firebreaks (such as human intervention), it enters the decay phase. In this stage, the fire intensity decreases, and flames diminish. Smoke and residual heat may still be present. Firefighters and equipment work during this phase to fully extinguish the fire and prevent rekindling.

9.3.2.
Special Events

While structural fires offer a diverse range of potential events, automotive fires typically follow a more gradual and predictable progression. Nevertheless, certain exceptional occurrences can take place during vehicle fires. This section delves into two of these outstanding events: backdraft and flashover. While this chapter will later explore the explosive hazards related to vehicle fires, it is essential to understand these exceptional events first.

9.3.2.1.
Backdraft

A backdraft is a dangerous occurrence that unfolds within confined spaces when a fire is starved of oxygen and suddenly replenished by an influx of fresh air. In the context of vehicle fires, this scenario can manifest within the cabin or trunk when flammable gases accumulate and reach a temperature where they are primed to ignite, but there is insufficient oxygen for combustion. The trigger for this precarious event often stems from the sudden shattering of a window or a door opening, permitting fresh air into the cabin. A loud, explosive sound and the forceful ejection of flames and smoke from openings often accompany a backdraft.

Also, the remaining window glasses might be broken and expelled, while doors or the trunk lid might burst open or blow out. While backdrafts in vehicle fires are not as frequent as in structural fires, it is essential for firefighters to be aware of the potential risks and to take appropriate safety measures when opening a trunk or preserved cabin during a vehicle fire.

9.3.2.2.
Flashover

Flashover, an event also more likely to happen in structural fires, can occur in a vehicle cabin fire when intense heat radiating from the initial fire source causes all the combustible materials inside the cabin to reach their ignition point almost simultaneously. As the temperatures escalate, the entire vehicle interior becomes engulfed in flames, resulting in a sudden and near-simultaneous ignition of the cabin materials. The confined space of the vehicle cabin can exacerbate the heat buildup, creating a thermal feedback loop that accelerates the fire growth, leading to flashover conditions. Flashover is a critical event that drastically changes the fire dynamics, and once it occurs, the fire becomes fully developed, generating extremely high temperatures, intense heat, and thick, dark smoke.

9.4.
Automotive Environment Peculiarities

9.4.1.
Vehicle Compartments

Even though the fire spread mechanisms seen in structural fires apply, several factors affect fire dynamics in a vehicle. To begin with, passenger vehicles have two or three compartments: an engine compartment, a cabin, and a trunk. The firewall, usually between the engine compartment and the cabin, will insulate both sections. It will reduce the speed at which a fire initiated in one of these sections will invade the other. While closed and in proper condition, doors, hoods, and windows will also limit the speed of the fire spread until a door is open (to retrieve occupants, for instance) or until glass windows shatter and fall apart. This would suddenly introduce fresh oxygen, boost the fire, and eventually trigger a backdraft.

9.4.2.
External Environment

Environmental conditions may have a significant impact: wind, rain, cold or hot weather, and so on, affecting the speed and direction in which flames propagate and the duration—or even a possible extinction—of the fire.

9.4.3.
Flammable Fluids

ICEVs carry significant amounts of fluids that can ignite at ambient temperatures, like gasoline, ethanol, and CNG. Diesel can ignite, if moderately heated, just above 150°F (~70°C). Moreover, several other fluids, like engine oil, transmission oil,

hydraulic brake fluid, engine coolant, and windshield washer fluids (which generally contain ethylene glycol), if warm enough, will also allow the fire triangle to start or maintain a fire tetrahedron running. Assuming, of course, there is a leakage on the related reservoir or pipelines. A leakage can be the initial cause (i.e., the root cause), allowing these combustibles to contact a suitable energy source and start the fire. In other instances, this leakage can be secondary damage, along with the evolution of the vehicle fire, releasing more combustible substances to the scene. If the fluid is under pressure, as gasoline at the input of fuel injectors, the leaky jet will carry flames far away.

9.4.4.
Plastic Content

Modern vehicles incorporate several hundred pounds of polymeric parts, contributing to a significant thermal load during a fire. Despite including flame-retardant additives in some plastic components, foams, and cabin upholstery, many are susceptible to ignition under the proper temperature or energy source. These polymers add fuel to the fire and release a substantial amount of energy, making contemporary car fires more challenging to combat.

In a fire, plastics within the cabin, including those in foams and upholstery, are prone to ignition. Frequently, these materials will also generate a significant amount of smoke and soot, explaining why the inner surface of the windows rapidly turns black.

On the other hand, thermoplastic parts (made of PVC, PA, PP, PE, or ABS, among others) inside the engine compartment usually do not contain flame retardants. Some might also melt down while carrying flames, propagating the fire to lower regions and causing a pool fire on the ground. This is also true with recent vehicles' bumpers and other plastic body parts.

9.4.5.
Low-Voltage Battery

The 12 V (or 24 V) lead–acid battery cannot be ignored as an energy source [9.2]. It may be tied to the root cause of the fire, acting as the electric energy source that heats up short circuits in the harness or overheats failed electro-electronic modules and parts. It may also allow secondary damages once these components are heated enough (in a fire initiated by a gasoline spill, for instance). Although relatively rare, this type of battery can also short circuit internally, being the root cause of the fire.

9.4.6.
Traction Battery

The traction battery of many xEVs is a lithium/liquid electrolyte battery variant. One crucial failure mode of this type is the internal short circuit, which usually leads to an abrupt rise in the temperature of the failed cell. Unfortunately, if heated far enough, the liquid electrolyte will degrade in exothermal reactions, further elevating the temperature of the entire cell in a process called thermal runaway. Conduction will transfer heat to surrounding cells, and these will also degrade. Meanwhile, many combustible and toxic gases and vapors will be created at high pressure and vented out of the cells/battery modules (creating a typical hissing sound) or deform and shatter their enclosures. The spread of short circuits along several cells or their connections will soon ignite these combustible gases. As stated by Sun et al., the propensity of self-ignition during regular charging, parking, and driving conditions due to the thermal runaway of the lithium battery makes xEV fires unique and very different from fires in ICEVs. In other words, fires involving a lithium traction battery represent a paradigm shift compared to typical ICEV fires [9.3].

9.4.7.
Fire Duration

Most vehicle fires develop and run faster than structural fires. If nobody acts, it will usually take less than 10 min to impose severe damages and less than 1 h to burn down the most combustible materials of a passenger car [9.4-9.7]. Meanwhile, nearby vehicles or constructions might be ignited in a domino effect.

Figure 9.4 depicts an experiment where the fire was purposely started in the engine compartment of a conventional passenger sedan using a sponge impregnated with gasoline. Maximum underhood temperatures surpassed 1650°F (900°C), while temperatures inside the cabin exceeded 1850°F (1000°C).

Figure 9.4 Fire in a passenger sedan.

| 1 min | 4 min 30 s | 6 min | 7 min 30 s | 9 min 30 s |
| 10 min 30 s | 10 min 40 s | 11 min | 13 min 30 s | 15 min |

Generally, steel parts will retain their forms because their melting points are above 2200°F (1200°C). However, parts made of aluminum and magnesium alloys—with melting points around 1200°F (650°C)—may melt down.

9.4.8.
Spread to Neighboring Vehicles

Car fires in expansive parking structures and large vehicle-transport vessels present unique challenges due to the proximity of unattended vehicles. If local fire suppression measures, such as sprinkler systems, are ineffective in containing the initial vehicle fire, the fire will likely spread rapidly to neighboring vehicles. This escalation poses a significant challenge for firefighting crews on arrival, given that the fire has already spread across multiple vehicles.

The close quarters and densely packed nature of parking structures and vehicle-transport vessels exacerbate the difficulty of extinguishing such fires. In large parking structures, accessibility issues may impede the firefighting process, potentially requiring more than a day to suppress the fire entirely. Similarly, fires in expansive car-transport vessels may necessitate an extended time frame, possibly exceeding a week, to bring the situation under control.

9.5.
Fires Initiated in the Engine Compartment

One of the most frequent causes of vehicle fires is fuel leakage in the engine compartment, reaching hot exhaust manifolds, catalytic converters, or turbocharger parts. Since the usual working temperature of their external surfaces is several hundred degrees Fahrenheit (up to ~500°C), an immediate ignition is likely if the fuel or its vapor contacts these. Also, when the engine is turned off, these parts will take several minutes to cool down, enlarging the window of opportunity for a fire to start on a recently idled (and faulty) vehicle.

Once the fire has started, the flame boundaries will enlarge radially inside the engine compartment, gradually burning plastics, hoses, and other combustible materials. Pipelines, connections, and reservoirs of the fuel or hydraulic systems will eventually be damaged, bringing these fluid combustibles to the scene. Harnesses, electronic modules, and parts will also be damaged, ultimately creating short circuits, augmenting temperatures, and further spreading the burning/damage zone.

The firewall will contain, for a while, the flames out of the cabin. However, if no countermeasures are taken, the flames will reach the lower part of the windshield and crack it due to the thermal shocks or the stresses imposed by different thermal expansions of the glass and its frame. The lower portion of the windshield will usually fall apart first, while the upper portion will remain attached to the frame—at least for a while. In certain vehicles, the flames might invade the cabin through the HVAC system or its ducts before the windshield shatters. As flames enter the cabin, they will advance through the plastic instrument panel, door and roof linings, seats, etc.

Figure 9.5, adapted from the NFPA 921 guideline, provides visual representations of the typical progressions of this scenario, offering both side and aerial views. Under the label "Evolution A," these images depict what had been conventionally considered the expected development of a fire originating in the engine compartment, with crucial evidence centered around fragments of the windshield, primarily attached to its upper frame. Conversely, the drawings labeled "Evolution B" present an alternative possibility. In modern vehicle designs featuring larger openings in the firewall surrounding the HVAC module and an increased reliance on plastic components, flames might access the cabin earlier, entering through ventilation ducts and potentially shattering the central or upper portion of the windshield before affecting its lower section.

Figure 9.5 Typical engine fire dynamics.

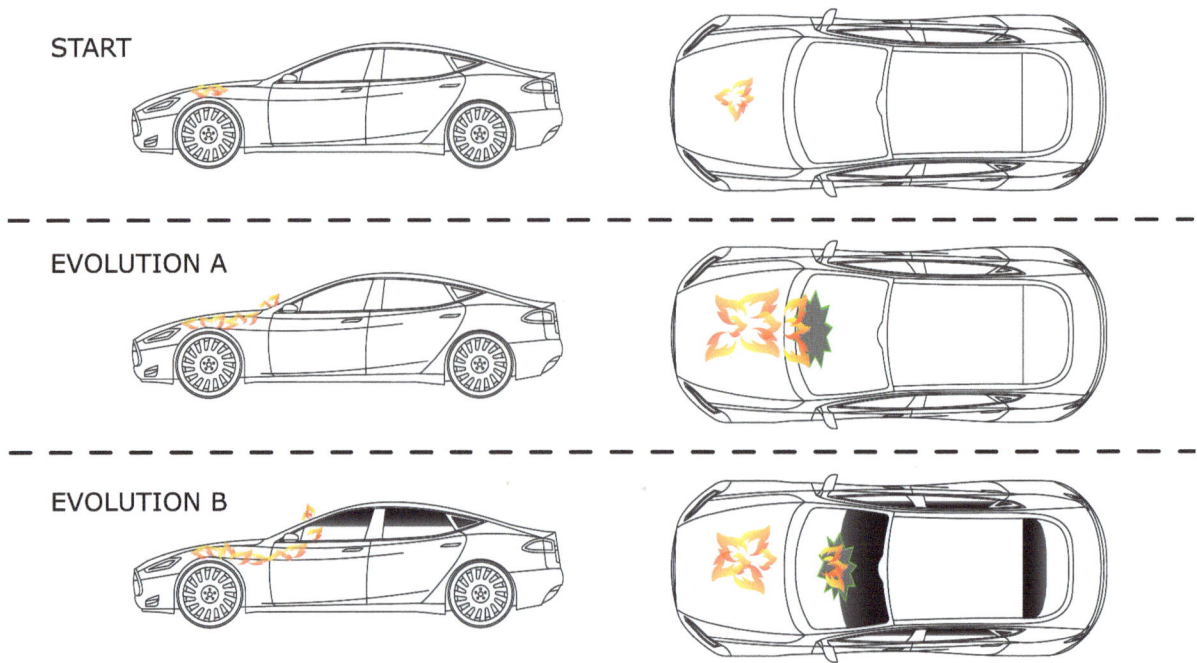

START

EVOLUTION A

EVOLUTION B

galimovna79, Maji Design/Shutterstock.com.

When the fire damage is not total, distinctive indicators of a fire originating in the engine compartment exist, which will be explored in greater depth in the Analysis section of this book. Typically, these signs include radial marks on the scorched paint of the hood or body and centered around the engine compartment. Depending on the vehicle's design and the location of the fire's origin, the upper part of the windshield may remain affixed to the metallic frame. At the same time, its lower section may disintegrate or even melt, as shown in **Figure 9.6**.

Figure 9.6 Fire initiated in the engine compartment.

Nadezda Mincheva/Shutterstock.com.

Without countermeasures, all the combustible materials will burn, and only metallic frames and parts will remain discernible. **Figure 9.7** provides a typical example of a vehicle burnt to the ground, made of steel chassis and body.

Figure 9.7 Scorched vehicle.

Tereshchenko Dmitry/Shutterstock.com.

9.6.
Fires Initiated in the Cabin

Fires originating in the passenger compartment exhibit distinct characteristics, setting them apart in their evolution. Consider the scenario where the ignition source is an accidental cigar ember that falls onto one of the seats shortly before the vehicle is parked and left unattended. When the seat upholstery catches fire, it ignites the structural foam beneath it, typically composed of polyurethane—a material that burns rapidly, releasing heat akin to a gasoline pool. As the flames intensify, adjacent components such as door and roof linings, carpets, upholstery, and other foams become additional fuel sources. Concurrently, plastic elements like the instrument panel may also ignite, producing a substantial volume of smoke and soot, which subsequently deposits on the inner surfaces of the window glass, darkening them.

With the accumulation of hot air in the cabin's upper portion, the windshield's upper region becomes subject to more significant stress, typically leading to its shattering before the lower section. As the upper windshield shatters, flames may escape through this opening, extending into the engine compartment and external areas. The flow of gases and vapors from this phenomenon draws fresh air and oxygen into the cabin, intensifying the progression of the fire. These dynamic sequences are illustrated in the upper diagrams of **Figure 9.8**.

Another scenario involves a fire initiated within the instrument panel, possibly due to an internal short circuit within a multimedia module. In this case, as depicted in the lower drawings of **Figure 9.8**, the central region of the windshield may be the first to succumb, leaving the glass borders still attached to its frame.

Figure 9.8 Cabin fire dynamics.

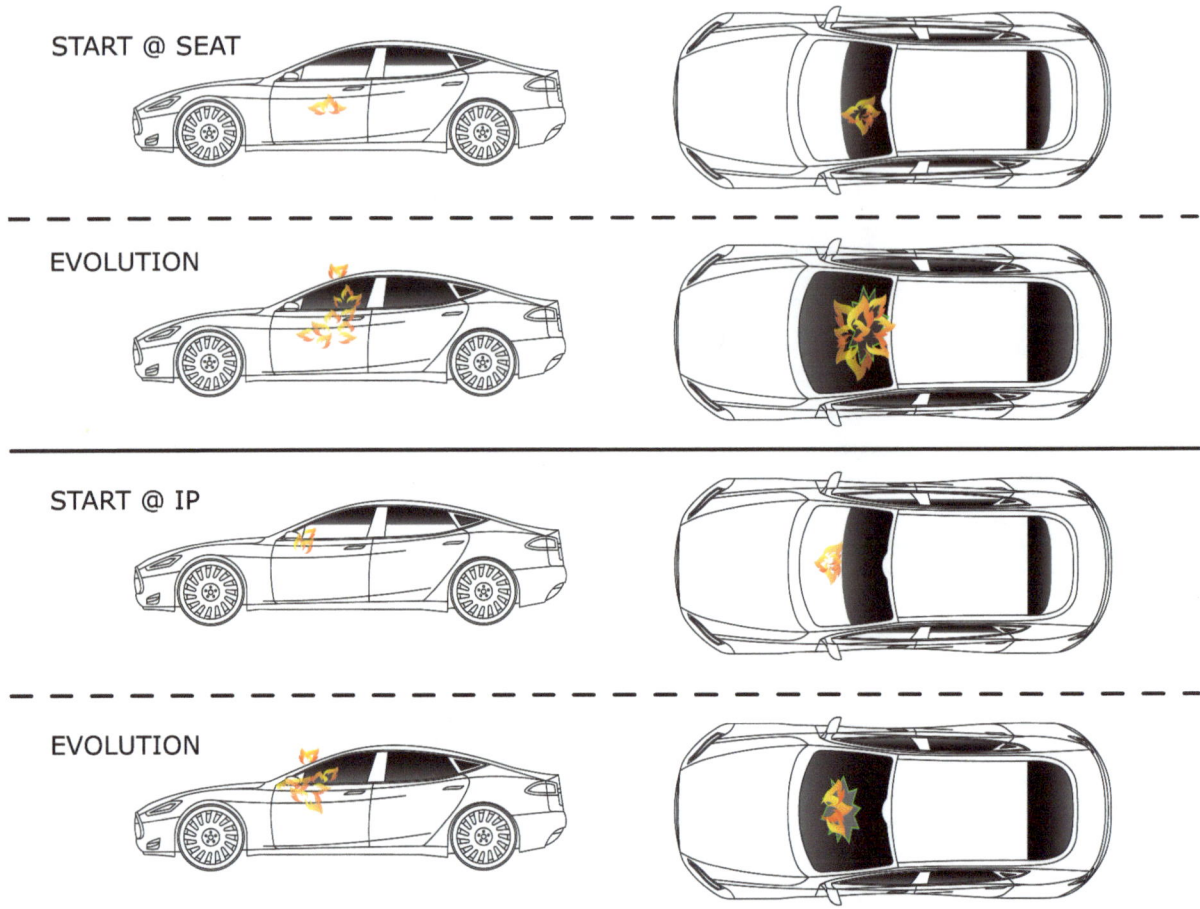

START @ SEAT

EVOLUTION

START @ IP

EVOLUTION

galimovma79/Shutterstock.com.

Figure 9.9 presents a characteristic illustration of a cabin fire scenario where the upper section of the windshield has become detached.

Figure 9.9 Cabin fire.

Dmitriy Prayzel/Shutterstock.com.

9.7.
Pool Fires

In the unfolding scenarios of Sections 9.5 and 9.6, releasing a substantial quantity of combustible liquids becomes a critical concern, often stemming from ruptured fuel tanks or pipelines due to flames or collisions. A "pool fire" could form directly beneath the vehicle, as depicted in **Figure 9.10**. It is important to note that most contemporary fuel tanks are made of plastic. When exposed to flames for an extended period, these tanks commonly rupture, forming a significant pool of fuel on the floor.

Figure 9.10 Pool fires.

(a) Engine Compartment

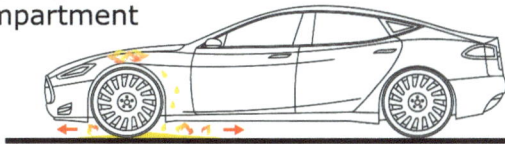

(b) Fuel Tank

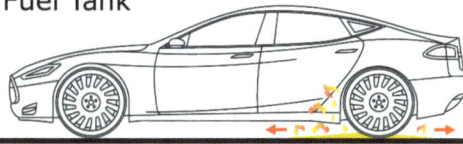

galimovma79/Shutterstock.com.

The pooling of combustible substances can rapidly propagate the fire along the lower body of the vehicle, engulfing tires and expanding in all directions. This also poses a severe hazard, particularly in car parking structures or vehicle transportation ships, where the floor may have a slope. Under such conditions, there is a heightened risk of a pool fire swiftly spreading the reach of the flames. This phenomenon, where flames spread through the fuel runaway, is often called a "running fuel fire." The eventual slope of the floor serves as a catalyst, accelerating the spread of the fire to multiple vehicles, intensifying the overall hazard, and complicating firefighting efforts.

9.8.
Fires Initiated in the Traction Battery

A dangerous failure mode is the thermal runaway for xEVs with a liquid electrolyte lithium traction battery. When this occurs, the electrolyte rapidly heats in a faulty cell. This leads to the degradation of various substances in the electrolyte through exothermic reactions, increasing the cell temperature even more, increasing its internal pressure, and generating toxic and combustible substances. At some point, the cell will vent or leak, or its casing will rupture, releasing these gases and vapors into the atmosphere and allowing them to burn.

Surrounding cells will progressively be overheated, expanding the damaged area and enlarging the burning envelope. A large volume of combustible gases escapes horizontally due to the obstacles created by the metal housing of the battery pack and by the metal floor of the vehicle. This frequently gives rise to expansive horizontal flames, reminiscent of a flamethrower, a phenomenon commonly referred to as jet fire, as presented in Chapter 8. These flames often reach other cars, buildings, and objects less than a few feet away (a couple of meters), further spreading the fire.

Moreover, recent xEVs more prominently use lightweight materials for the body (aluminum and polymers). The high temperatures of the combustion of the substances released by the damaged batteries rapidly melt or consume the vehicle's body, leaving a low-profile skeleton. **Figure 9.11** depicts an example, while **Figure 9.12** sketches the complete evolution of this type of fire.

A crucial facet of addressing traction battery fires lies in the significant volume of water required for effective extinguishment. Per NFPA guidelines, tackling a fire in an xEV may necessitate more than 2600 gal (10,000 L) of water—an order of magnitude higher than the typical amount required for ICEV fires. Adding to the complexity, conventional firefighting methods may prolong the extinguishing process, often spanning several hours.

Figure 9.11 BEV body consumed by fire.

Javier ki/Shutterstock.com.

Figure 9.12 Traction battery fire evolution.

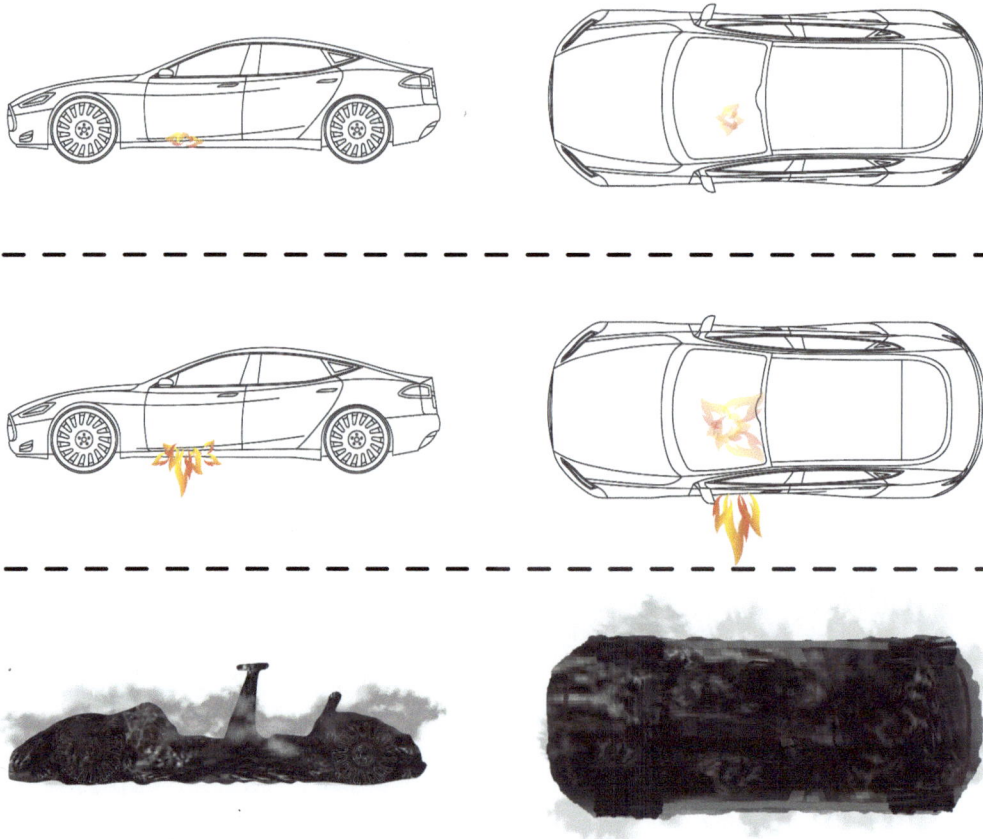

galimovma79, Mind Pixell/Shutterstock.com.

These disparities stem from the thermal degradation of various substances within the LIBs, leading to the generation of oxidizers. Additionally, the metallic enclosures surrounding battery modules present dual challenges: hindering direct contact with external firefighting water and providing thermal insulation to the battery cells, complicating their cooling during firefighting efforts. This intricate interplay of factors underscores the need for specialized strategies and heightened awareness when dealing with traction battery fires.

Another noteworthy concern within the dynamics of xEV fires is the potential for rekindling after initial suppression. When a significant quantity of electric energy remains stored within the traction battery, commonly referred to as "stranded energy," the risk arises that leakage currents or new short circuits could trigger a fresh thermal runaway event in the remaining battery cells. Incidents of this nature have been documented in various vehicles, as examined by Bisschop et al. [9.8] and Sun et al. [9.9], sometimes occurring within mere hours or even days following the initial fire suppression. These intricate aspects of xEV fire safety are revisited in other chapters of this book.

9.9.
Burn Patterns

In the analysis of vehicle fires, specific visual patterns are considered relevant and frequent as they can provide valuable insights into the critical information about the fire. Fire investigators often study these burn patterns to determine the point of origin of the fire, the eventual use of fire accelerants, the sequence of events leading to the fire, and its path or progression. Some of these patterns are described in more detail as follows [9.10-9.12].

9.9.1.
Seat of Fire

The seat of fire refers to the specific area where the fire originated. More severe damage often characterizes it and can be a critical clue in determining the cause of the fire. Of course, it is more discernible if the fire was extinguished before a large area was damaged.

Figure 9.13 depicts the damage on a semitrailer caused by a fire initiated by having the parking brake applied while the vehicle was being driven. The heat-and-burn pattern is above and around the tires on the rear axle.

Figure 9.13 Seat of fire on a semitrailer.

Reprinted from SAE Technical Paper 2013-01-0207. © SAE International.

9.9.2.
V Pattern

A V-shaped pattern on vertical surfaces is a critical indicator of the fire's path and is a well-known clue for structural fire investigators. Typically, the apex of the V points toward the origin of the fire, aiding investigators in pinpointing the initial ignition source. While less commonly observed in vehicles, it can be detectable under specific circumstances on internal or external vertical surfaces, such as doors and lateral panels of larger vehicles.

9.9.3.
Charring

Charring refers to the visible blackening and burning of materials, such as plastic or other solid combustibles. The extent and depth of the charring can provide clues about the duration and intensity of the fire. **Figure 9.14** portrays a largely charred instrument panel.

Figure 9.14 Charred instrument panel.

travelarium.ph/Shutterstock.com.

9.9.4.
Melting and Dripping

Melting and dripping patterns on surfaces or materials can indicate the origin of the heat source and can also point to the presence of accelerants or other substances that affect the behavior of the fire. **Figure 9.15** exemplifies a melted rear lantern caused by a fire in the building next to it.

Figure 9.15 Melted rear lantern.

Nelson Antoine/Shutterstock.com.

Pouring a liquid accelerant, for example, gasoline, over a vehicle and lighting it will likely result in dripping of the excess gasoline along several parts, causing superficial damage, as the flames appear, on relatively distant parts. The characteristic burn marks of a dripping path might survive if the vehicle fire is interrupted at an early stage, as can be inferred in **Figure 9.16**.

Figure 9.16 Burn patterns suggestive of arson.

Denis Mamin/Shutterstock.com.

9.9.5.
Damage to Wiring and Electric Components

Assessing the state of vehicle wiring and components is instrumental in uncovering electrical anomalies that could have triggered the fire. Detecting signs of electrical arcing, such as the presence of copper beads, charred wires, or damaged electrical components, can point to electrical faults as a possible ignition source. Chapters 4, 5, and 6 of this book have already presented some failure and damage mechanisms.

The forthcoming chapters will further explore scenarios in which these damages may not be the root cause of the fire but a consequential outcome.

9.9.6.
Flow Path

Once started, a vehicle fire will spread along various fuel sources, such as fuel lines, seats, and interior components. When the vehicle's damage is not complete, a path of destruction is occasionally discernible, showing how the fire traveled within the car. This information helps investigators determine potential ignition sources and what materials were involved. Also, abnormal paths might indicate an intentional fire.

9.9.7.
Other Burn Patterns

Vehicle fires can often display distinct patterns influenced by the vehicle's design, including factors like the engine's location, the fuel tank's placement, and the unique geometries of various body parts. Additionally, specific patterns can be observed along openings and doors, providing valuable insights into their status during the fire incident. For instance, **Figure 9.17** shows a vertical mark along the right edge of the driver's door. This mark indicates that the door was open, at least during a part of the fire.

Figure 9.17 Door open during fire.

Reprinted from SAE Technical Paper 2006-01-0548.
© SAE International.

Conversely, sizable areas on vehicle body parts, such as doors and hoods, often exhibit diverse and multicolored spots. Interpreting these marks can be challenging, as there might be multiple reasons behind their contrasting appearances.

One factor contributing to these differences is the varying degrees of combustion among the multiple layers of paint. Additionally, the range and duration of high temperatures experienced during the fire create these distinctive patterns. The evolution of these temperatures can result from structural attachments and reinforcements located on the inner side of sheet metal components, as evidenced by the geometric patterns observed on the hood in **Figure 9.18**.

Figure 9.18 Geometric burn patterns on the hood.

Furthermore, when the protective paint layers on steel components burn away, these parts undergo high-temperature oxidation and become susceptible to atmospheric corrosion. These conditions lead to the formation of distinct oxide layers, including wustite (FeO), magnetite (Fe_3O_4), and hematite (Fe_2O_3). These oxide layers continue to evolve in terms of thickness, three-dimensional structure, and color even after extinguishing the fire.

For a more in-depth analysis of this phenomenon, one can turn to the research

conducted by Colwell and Babic [9.8]. **Figure 9.19** offers a practical example illustrating a wide range of colors and shades resulting from the high-temperature oxidation of the steel body and subsequent atmospheric corrosion.

Figure 9.19 Several rust colors and shades.

9.9.8.
Vehicle Components Affected

Examining the specific components of the vehicle severely damaged or entirely consumed by the fire can yield valuable insights into the origin, intensity, and progression of the fire. For example, aluminum alloy wheels might melt down if exposed to a fire pool along the incident, as the image depicted in **Figure 9.20** seems to indicate.

Also, in cars equipped with plastic fuel tanks and aluminum alloy wheels, the wheel closest to the fuel tank's opening may melt, while the other wheels do not reach the melting temperature. Typically, this outcome arises from the plastic fuel filler pipe compromise during the fire progression, resulting in gasoline leakage around the nearest wheel. Consequently, this scenario raises the local temperature and prolongs the duration of high heat exposure around that particular wheel.

Figure 9.20 Wheel melt down.

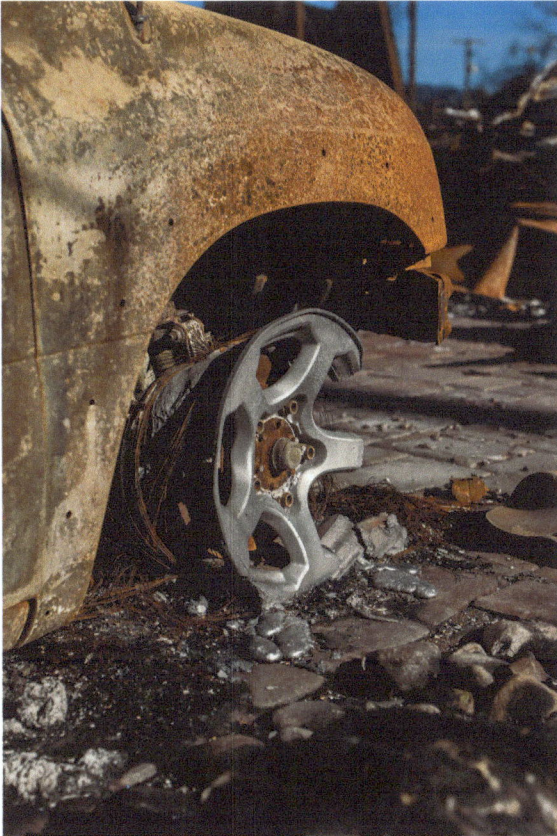

Vlad Teodor/Shutterstock.com.

9.9.9.
Multiple Origins

In certain instances, vehicle fires may exhibit multiple points of origin, suggesting the presence of either numerous ignition sources or deliberate actions.

It is important to note that each vehicle fire is unique, and several burn patterns might be present, making the analysis complex and requiring expertise in fire investigation. Analyzing these burn patterns with other investigative techniques allows experts to reconstruct the sequence of events leading to the

fire and help determine the most likely cause of the incident, whether it be a mechanical failure, electrical issue, arson, or any other factor. This information is crucial for insurance purposes, legal investigations, and vehicle fire safety improvements.

Figure 9.21, sourced from a paper authored by Shields and Scheibe [9.9], offers an illustrative case where the central hood area's paint is charred, along with the paint on the front corner of the fender. In this particular instance, no internal damage accounted for the transfer of heat to this point, leading to the determination that arson was the root cause of the fire.

Figure 9.21 Two separated burn patterns.

Reprinted from SAE Technical Paper 2006-01-0548.
© SAE International.

9.10.
Wind Influence

Wind can significantly interfere with the direction of the flames and smoke, as demonstrated in **Figure 9.22** for a parked car. If the vehicle is in motion, consider the influence of the relative movement of the air around it [9.13-9.15].

Figure 9.22 Wind deflecting flames and smoke.

Indeed, the influence of wind on a vehicle fire is multifaceted, with the potential to alter its progression in numerous ways, including its spread, intensity, and overall behavior. The impact of wind on a vehicle fire is contingent on various factors, encompassing its direction, velocity, the surrounding environment's configuration, and the specific area of the vehicle where the fire is unfolding. The ensuing sections outline several ways in which wind can shape the course of a vehicle fire.

9.10.1.
Accelerating Fire Spread

Wind can accelerate the spread of a vehicle fire by providing a continuous and abundant supply of oxygen to the flames. Oxygen is essential for combustion, and when the wind blows, it pushes fresh air toward the fire, enabling it to burn more intensely and spread more rapidly.

9.10.2.
Fire Direction and Path

The direction of the wind can dictate the path of the fire. If the wind blows directly toward the vehicle fire, it can push the flames in a particular direction, potentially affecting nearby structures, vegetation, or other vehicles. However, wind can be beneficial in some circumstances if it drives the flames from a burning engine compartment away from the vehicle cabin, for instance.

9.10.3.
Increased Heat Transfer

Wind can increase the heat transfer rate from the fire to surrounding objects and surfaces. By blowing hot gases and flames, wind can cause nearby objects to ignite, contributing to the spread of the fire.

9.10.4.
Smoke Spread and Visibility

Wind can disperse smoke and gases the fire produces over a wider area. This can reduce visibility for firefighters and bystanders, making assessing and responding effectively more challenging.

9.10.5.
Ember Transport

Embers and flaming debris, propelled by the wind, can travel considerable distances, potentially kindling spot fires in locations far removed from the initial fire source. This phenomenon introduces added complexities to firefighting endeavors and contributes to an expanded fire perimeter. Furthermore, suspended embers from forest or roadside fires may eventually infiltrate passing vehicles, gaining access to their engine, cabin, or cargo compartments, thereby instigating car fires.

9.10.6.
Fire Control and Suppression

Wind can hamper firefighting efforts by spreading the fire beyond the initial point of origin. Sudden changes in wind intensity or direction may also make controlling the perimeter of the fire difficult, making containment more challenging. It can also create fire whirls or vortices, causing localized areas of intense burning and spreading fire in unexpected directions.

Therefore, wind can affect the safety of firefighting operations, the effectiveness of suppression efforts, and the potential for fire to spread to nearby structures or fuel sources. Firefighting personnel must be aware of wind conditions and adapt their tactics to manage and control vehicle fires effectively.

9.11.
Cars Always Explode in Movies

The recurrent appearance of fantastic vehicle explosions in thriller movies—apart from the emotions pursued by their directors—has little correlation with real life. In cinema, a minor collision, or perhaps the puncture of a bullet, quickly makes the fuel tank and the entire vehicle burst into flames.

From a scientific perspective, usually, there is no reason for a liquid fuel tank to explode in such a way. Even though it might be full of gasoline, there is no oxygen inside, or not at an adequate proportion with the fuel vapors, to allow combustion. The combustion will happen only externally to the tank, as the gasoline or its vapor contacts fresh air, provided an ignition source or a fire tetrahedron is already in place. Therefore, it usually results in a relatively slow burn process. If the tank breaks and suddenly releases substantial fuel, giant flames will result—yet no "Hollywood car explosion" in the general case.

There are exceptions, of course. Imagine a tank filled with LPG and a rusted metallic case subjected to excessive pressure. At some point, the enclosure will break apart and the tank will explode. This may or may not create flames or cause the gas to burn simultaneously.

Another example is a truck towing a large combustible liquid tank, moving too fast along a highway curve. If the set overturns, the impact might open one of the tank lids (or fracture the tank), while the friction between the steel case and the asphalt pavement will create multiple sparks. *Voilà!* the liquid is ignited while being spilled along the highway, leaving a devastation similar to an actual explosion [9.16].

9.12.
Parts That Might Explode

With a fire already in progress, some vehicle parts and modules can explode, possibly injuring occupants and rescue teams. Here, the role of tires, airbags, pretensioners, gas struts, and gas tanks (LPG, CNG, and hydrogen) will be explored.

Although the tires are filled with air (or eventually nitrogen), which is not combustible, as the flames around the tire heat the rubber walls, these will begin to degrade. At the same time, the temperature rise will increase the internal pressure. At some point, the tire wall will rupture, often hurling high-speed fragments around it, potentially injuring people in the vicinity, and launching burning pieces (that can reach other fuels). Also, the sudden expansion of the air inside the tire will displace and carry existing flames far away. The risks are higher for larger tires (e.g., buses and heavy trucks).

Airbag gas generators are parts that also deserve due care and attention. Usually, they are filled with sodium azide (NaN_3), a relatively unstable substance, necessary to generate a large amount of nitrogen gas to quickly inflate the airbag during a collision. To achieve that, the control module delivers an electrical discharge, triggering the decomposition of this substance. However, if the cabin of a parked car is engulfed in flames, the sodium azide reservoir will be overheated, triggering the same reaction and exploding. But in this situation, the fire may have already damaged the fastening parts and surrounding covers. This creates the potential for the explosion to project dangerous fragments around, a situation that firefighters and first responders must be well aware [9.17].

The pretensioners of seat belts, usually containing a small amount of explosive, can behave similarly when heated, creating risks for eventual occupants and rescue teams. Compared to airbag gas generators, pretensioners produce a much smaller amount of gas, presenting a somewhat lower risk during a cabin fire. **Figure 9.23** pictures one of these modules (the white cylinder), incorporated within the buckling device. It might also be located in the lower region of B and C pillars, as part of the seatbelt retraction mechanism.

Figure 9.23 Explosive (gas generator) in a seatbelt pretensioner.

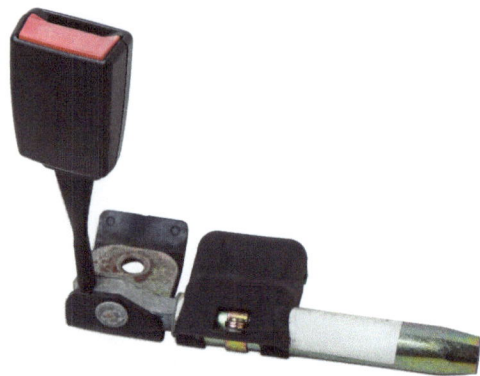

Aleksandr Kondratov/Shutterstock.com.

Gas-filled struts can also be damaged along a fire, whether used in the suspension system or to help keep hoods open. The exposure to excessive temperatures might increase the internal pressure to a point where their tubular case might burst or just deform the end cap enough so that the internal piston or rod is expelled with violence, posing another risk. **Figure 9.24** shows a typical gas-filled hood strut, while **Figure 9.25** exhibits one that burst during a vehicle fire.

Figure 9.24 Gas-filled strut.

Hanjo Stier/Shutterstock.com.

Figure 9.25 Hood strut burst during fire.

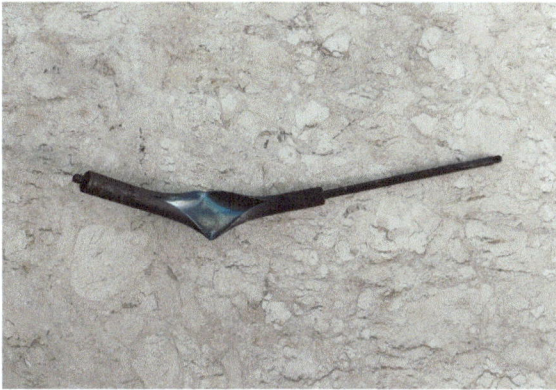

© SAE International.

A tank containing LPG, CNG, or hydrogen might explode. Usually, not because a combustion happens inside the tank, but because the tank case fractures or its connections leak under abnormal circumstances: collision, excessive pressure, fatigue, corrosion, buildup of internal pressure caused by flames external to the tank, etc.

Finally, the vehicle cabin faces a potential risk of explosion, should flammable gases amass within its confines. This peril may arise from a gas-propelled vehicle experiencing a leakage, improper handling of gas containers (such as

aerosol deodorant cans), or combustible gases emanating from an LIB undergoing thermal runaway.

9.13.
Cause: Chemical ➜ Electrical or Vice Versa?

The root cause of the example presented in Section 9.5 is usually named a "chemical" or "thermal" event, that is, a combustible fluid is heated up—or meets a hot surface—reaching a temperature that ignites the combustion process. Another possibility for the root cause is an "electric" event, as discussed in the second example of Section 9.6. A short circuit inside the instrument panel was considered the event that started the fire. The reader was purposely driven to assume these root causes as a basis to under-stand the resulting fire dynamics.

But, in most real-life situations, the fire investigator will not have upfront information regarding the root cause. And in some cases, the level of destruction of the vehicle might provide traces suggesting both possibilities, that is, a chemical fire that, along its evolution, created short circuits and left evidence that could be improperly interpreted as an electric fire; also, an electric fire might create flames that will, at some point, damage fuel hoses and leave traces that could be misinterpreted regarding the actual root cause.

For now, keep in mind that, at first glance, the analysis of a burnt vehicle may resemble the chicken-and-egg problem: which was the actual root cause: the burning of flammable substances or the electrical event? Chapters 13 and 14 will pursue the clarification of this dilemma.

References

[9.1]. National Fire Protection Association, "Guide for Fire and Explosion Investigations—NFPA 921," 2021.

[9.2]. Barnett, G., *Vehicle Battery Fires: Why They Happen and How They Happen*, SAE International Book R-443 (Warrendale: SAE International, 2017, ISBN:978-0-7680-8143-5.

[9.3]. Larsson, F., Andersson, P., and Mellander, B., "Battery Aspects on Fires in Electrified Vehicles," in *3rd International Conference on Fire in Vehicles*, Berlin, Germany, October 1–2, 2014, ISBN:978-91-87461-87-3.

[9.4]. Colwell, J., "Full-Scale Burn Test of a 2001 Full-Size Pickup Truck," *SAE Int. J. Trans. Safety* 1, no. 2 (2013): 450-466, doi:https://doi.org/10.4271/2013-01-0214.

[9.5]. DeMarois, P.H., Ballard, W., Engle, J., West, G. et al., "Full Scale Burn Demonstration of Two 2013 Ford Fusions - Arc Mapping Analysis," SAE Technical Paper 2018-01-1439 (2018), doi:https://doi.org/10.4271/2018-01-1439.

[9.6]. Jiang, X., Zhu, G., Zhu, H., and Li, D., "Full-Scale Experimental Study of Fire Spread Behavior of Cars," in *2017 8th International Conference on Fire Science and Fire Protection Engineering*, Nanjing, China, 2017, https://doi.org/10.1016/j.proeng.2017.12.016.

[9.7]. Li, D., Zhu, G., Zhu, H., Yu, Z. et al., "Flame Spread and Smoke Temperature of Full-Scale Fire Test of Car Fire," *Case Studies in Thermal Engineering* 10 (2017): 315-324, doi:https://doi.org/10.1016/j.csite.2017.08.001.

[9.8]. Bisschop, R., Willstrand, O., Amon, F., and Rosengren, M., "Fire Safety of Lithium-Ion Batteries in Road Vehicles," RISE Research Institutes of Sweden, 2019, https://doi.org/10.13140/RG.2.2.18738.15049.

[9.9]. Sun, P., Bisschop, R., Niu, H., and Huang, X., "A Review of Battery Fires in Electric Vehicles," *Fire Technology* 56 (2020): 1361-1410, doi:https://doi.org/10.1007/s10694-019-00944-3.

[9.10]. Colwell, J. and Babic, D., "A Review of Oxidation on Steel Surfaces in the Context of Fire Investigations," *SAE Int. J. Passeng. Cars - Mech. Syst.* 5, no. 2 (2012): 1002-1015, doi:https://doi.org/10.4271/2012-01-0990.

[9.11]. Shields, L. and Scheibe, R., "Computer-Based Training in Vehicle Fire Investigation Part 2: Fuel Sources and Burn Patterns," SAE Technical Paper 2006-01-0548 (2006), doi:https://doi.org/10.4271/2006-01-0548.

[9.12]. Smith, N., Hicks, W., Gorbett, G., Hopkins, R. et al., "Vehicle Fire Burn Pattern Study," in *International Symposium on Fire Investigation Science and Technology, ISFI 2010*, Sarasota, FL, 2010.

[9.13]. Engle, J.J., Buckman, J.L., Williams, J., Kemnitz, E. et al., "An Analysis of the Effects of Ventilation on Burn Patterns Resulting from Passenger Compartment Interior Fires," SAE Technical Paper 2020-01-0923 (2020), doi:https://doi.org/10.4271/2020-01-0923.

[9.14]. Leffert, M., "Effects of Wind Speed and Longitudinal Direction on Fire Patterns from a Vehicle Fire in a Compact Car," SAE Technical Paper 2017-01-1353 (2017), doi:https://doi.org/10.4271/2017-01-1353.

[9.15]. Zhu, H., Gao, Y., and Guo, H., "Experimental Investigation of Burning Behavior of a Running Vehicle," *Case Studies in Thermal Engineering* 22 (2020): 100795, doi:https://doi.org/10.1016/j.csite.2020.100795.

[9.16]. Fox 2 Detroit, "Video Shows Tanker Truck Crashing on I-75, Explodes," July 13, 2021, accessed February 2024, https://www.youtube.com/watch?v=hpooPpULjCg.

[9.17]. Morningstaric "Car Explosion in Los Angeles 19th August 2011 Brave Fire Fighter," August 20, 2011, accessed May 2024, https://www.youtube.com/watch?v=zLoF99LYdts.

Section 2: Prevention

Chapter 10

Product Design

10.1.
Introduction

In more than a century of the existence of the automotive industry, the pursuit of robust designs remains paramount. A robust design ensures that a vehicle performs as expected under a wide array of conditions and upholds the highest standards of quality and reliability. This pursuit becomes even more critical when considering the potential consequences of design shortcomings—particularly when they may lead to fire incidents.

Robust design encompasses the meticulous selection of components and materials, guided by precise specifications and detailed 3D renderings. These components are engineered to withstand real-world stresses with substantial safety margins, assuring their durability and resilience.

However, the pursuit of robustness is not always straightforward. Technical and cost constraints can occasionally impede the attainment of the desired level of robustness. In such cases, it becomes imperative to implement additional safeguards in the design. These barriers are strategically positioned to prevent failures, curtail their propagation, or, at the very least, diminish the speed at which other components may be adversely affected.

This chapter briefly reviews the evolution of product design methodologies, explores the fundamental aspects of robust designs, acknowledges the relevance of technical specifications, and delves into the crucial roles of mechanical, electrical, and logic (software) barriers. Additionally, it considers the importance of designing safety layers for xEV batteries.

10.2.
Product Design Strategies

Over the decades, the automotive industry has undergone a remarkable transformation in its approach to product design, driven by a relentless pursuit of performance, safety, quality, mass production, cost reduction, waste reduction, and innovation. This evolution has benefited from, or yielded, several key design strategies and methodologies, each contributing to robust and reliable vehicles that meet their customers' needs and desires [10.1-10.3].

10.2.1.
Early Age and Conformance

In its nascent stage, automotive design was primarily utilitarian and focused on basic functionality. As the industry matured, there was a growing emphasis on conformance to established safety and quality standards.

10.2.2.
Zero Defects and Continuous Improvement

Post-World War II, the "zero defects" philosophy and continuous improvement methodologies like total quality management (TQM) gained prominence. These approaches aim to minimize defects and enhance product reliability, prioritizing quality control and reliability in manufacturing processes.

10.2.3.
Precision Engineering and Six Sigma

Precision engineering has been integral to the automotive industry since its inception. It involves using ever-evolving technologies and techniques to achieve higher accuracy, reliability, and performance in designing and manufacturing high-precision components and systems. Around the turn of the century, several players in this industry also adopted Six Sigma. This data-driven methodology provides tools and techniques to define, measure, analyze, improve, and control business processes. It aims to minimize variations and defects in design and manufacturing processes, leading to higher-quality products with fewer flaws.

10.2.4.
Computer-Aided Design (CAD)

The advent of CAD revolutionized automotive design in several aspects, enabling precise and efficient planning. CAD has replaced manual drawing boards and has made it possible to create 3D models of cars, which can be easily modified and tested for aerodynamics, safety, and numerous other factors. This has led to a significant reduction in the time and cost required for designing cars.

10.2.5.
Integrated Product Development (IPD)

IPD strategies emerged, emphasizing the integration of various aspects of automotive design, from engineering to manufacturing and marketing. This holistic approach ensured design elements aligned with safety, regulatory standards, reliability, and market demands.

10.2.6.
Sustainability and Efficiency

In recent years, automotive design has shifted toward sustainability and fuel efficiency. Innovations in lightweight materials, aerodynamics, and hybrid/electric vehicle technologies have improved overall vehicle efficiency and reduced emissions.

10.2.7.
Advanced Technologies

Rapid advancements in autonomous driving, connectivity, and electrification mark the current era. While introducing new challenges, these innovations offer opportunities to enhance vehicle safety through advanced safety systems, real-time diagnostics, and connectivity.

10.3.
Robust Design

Robust product design is essential in the ongoing efforts of the automotive industry to reduce failures and safeguard the well-being of vehicle occupants and the broader public. This holistic approach to design encompasses several facets that, when combined, enhance the overall quality, reduce failures, and contribute significantly to safety and fire prevention.

From a statistical point of view, a robust design can be seen as a design with reduced variability around a target performance. Given a large population of manufactured products, higher variability would imply that many parts will exhibit poor performance, that is, performance below an acceptable limit, represented by the red area on the chart of **Figure 10.1**. In practice, these specimens will be more likely to fail. The blue curve shows an improved design in which the performance variation has been reduced. Therefore, the number of failures of a given production volume will be diminished, as represented by the smaller blue area on the left.

Figure 10.1 Design improvement from a statistical perspective.

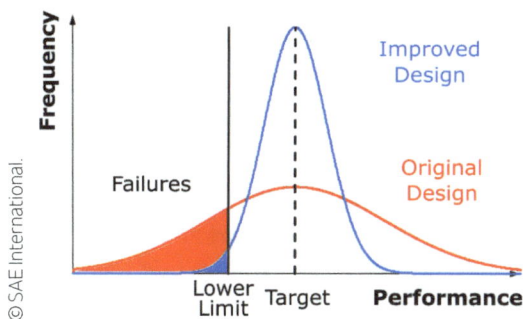

One fundamental aspect of robust design involves the selection of materials. These materials should be durable and cost-effective, but they must also possess fire-resistant qualities adequate to the specific part. The correct choice in the design phase significantly reduces the risk of fires or limits their propagation in the event of an incident.

Robust design is also paramount in the electrical system. An intricately designed electrical system should incorporate safeguards like fuses, overload protection, and proper insulation to thwart short circuits and electrical faults that could lead to fires.

Proper design and positioning are crucial when considering fuel systems. Robust designs ensure that components like fuel tanks, lines, and connectors are secure and equipped with safety features to mitigate fuel leaks following a collision.

To prevent leakages of other fluids, robust designs ensure the secure sealing and routing of fluid systems, such as brake lines, hydraulic systems, and coolant systems. This is crucial to prevent leaks that could lead to fires if these fluids come into contact with hot engine components.

Effective heat management is essential in the engine compartment, a high-temperature environment. Robust designs incorporate heat shields, cooling systems, and the appropriate arrangement of components to prevent overheating and reduce the risk of fires caused by temperature spikes.

Maintaining the integrity of the exhaust system is also a priority. A well-designed exhaust system prevents the escape of hot gases while limiting the heat transfer between vehicle parts, which could otherwise ignite nearby flammable materials.

Crashworthiness is yet another facet of automotive safety. Vehicles must be engineered to withstand collisions without causing fires. This involves careful structural engineering, including crumple zones and impact-absorbing materials, to protect, for instance, fuel systems from damage and, therefore, prevent fire initiation during accidents.

The rise of xEVs places additional emphasis on robust design. The traction battery system requires meticulous engineering, as seen in more detail in upcoming topics. This includes thermal management systems, crash-resistant battery enclosures, and redundant safety mechanisms to forestall thermal runaway and fires in the event of accidents or malfunctions.

Moreover, compliance with safety standards and regulations is integral to robust product design. Meeting or surpassing several of these standards is vital for minimizing the risk of car fires, besides ensuring overall vehicle safety [10.4, 10.5].

In conclusion, robust product design serves as a linchpin in mitigating car fires by addressing vulnerabilities in materials, electrical systems, fuel systems, and other components. It bolsters vehicle safety and contributes to the broader protection of road users. Vehicle manufacturers must continually innovate and refine their designs to stay ahead of emerging risks while automotive technology evolves.

10.4.
Specifications and Regulations

Technical specifications play a vital role in the automotive industry for several reasons. They were introduced to standardize various vehicle design and manufacturing aspects, ensuring compatibility, enabling mass production, avoiding failures, and reducing fire risks.

One critical function of technical specifications is facilitating compatibility between different vehicle components. They ensure that all parts are meticulously designed to work seamlessly together, ensuring the proper functioning of the vehicle. This compatibility is crucial for manufacturing vehicles efficiently and cost-effectively.

Furthermore, technical specifications are instrumental in mass production. By stipulating uniform standards for materials, dimensions, assembly procedures, and quality control, these specifications reduce the likelihood of defects, ultimately increasing efficiency on the assembly line.

Technical specifications also serve as preventive measures to avoid failures. They ensure that all vehicle parts are designed to withstand the stresses and strains of everyday use, reducing the chances of premature failures, which can be hazardous and costly.

In addition to failure avoidance, technical specifications contribute to fire risk reduction. They achieve this by mandating materials less likely to ignite or burn and ensuring proper insulation and protection of critical components. And, as seen in other topics, since many fires result from failures, avoiding failures in general also contributes to preventing fires.

This book has no intention to cover the ever-evolving set of specifications and regulations. Yet, here are some technical specification categories that contribute to fire prevention, either directly or through the reduction of failures that might evolve into fires [10.6-10.15]:

- Flammability.
- Resistance to vibration, mechanical forces, and impacts.
- Collision resistance at a vehicle level.
- Resistance to extreme temperatures and thermal shocks.
- Resistance to electric overloads and transients.
- Sealing/resistance to solid particles and dust ingress.
- Sealing/resistance to water ingress.
- Resistance to humidity.
- Resistance to automotive fluids.
- Resistance to a saline environment.
- Vehicle resistance to all water manifestations: tropical rain, snow, clouds created by nearby vehicles, car washing, passage through flooded areas, etc.
- Resistance to ultraviolet rays.
- Resistance to assembly errors.
- Resistance to customer abuse.
- Resistance to cycles in virtually all previous categories.
- Endurance/performance over prolonged use and exposure.

While most automotive industry manufacturers have specifications, several countries and global regions have specific regulations that vehicles must obey to be commercialized and driven there. At least the following FMVSS, applicable to the US, have a clear relationship with fire prevention and mitigation [10.16-10.20]:

- FMVSS 301: Fuel System Integrity.
- FMVSS 302: Flammability of Interior Materials.

- FMVSS 303: Fuel System Integrity of Compressed Natural Gas Vehicles.
- FMVSS 304: Compressed Natural Gas Fuel Container Integrity.
- FMVSS 305: Electric-Powered Vehicles: Electrolyte Spillage and Electrical Shock Protection.

Similarly, at least the following European Commission regulations applicable to the European Union have a similar focus [10.21-10.25]:

- Prevention of fire risks (liquid fuel tanks): EC No. 2019/2144, UN R34.
- Battery electric vehicles safety: EC No. 2019/2144, UN R100.
- CNG systems: EC No. 2019/2144, UN R110.
- Fire resistance of interior materials: EC No. 2019/2144, UN R118.

10.5.
Other Prevention Tools

The automotive industry applies many theoretical tools, documents, analyses, and practices that, besides contributing to reducing failures, help reduce the number of fire incidents. Here is another nonexhaustive list:

- Best practices.
- Computational fluid dynamics (CFD).
- Control plans.
- Design failure mode and effect analysis (DFMEA).
- Field incident analysis and feedback.
- Lessons learned.

- Process failure mode and effect analysis (PFMEA).
- Recalls.
- Risk identification and mitigation.
- Root cause analysis and prevention/correction tools.
- Six Sigma tools.
- Thermal mapping.

10.6.
Mechanical Barriers

Several parts and assembly precautions are employed to reduce exposure to mechanical stresses that could otherwise impair function-ality or allow failures. Here is a range of elements that suit this category:

- Vibration suppression: Using cushions, mounts, clips, clamps, fasteners, and other damping materials, like rubbers, foams, and tapes. **Figure 10.2** shows some examples of rubber suspension parts.

Figure 10.2 Rubber suspension parts.

Aumm graphixphoto/Shutterstock.com.

- Abrasion resistance: Corrugated plastic tubes are often used in automotive wiring harnesses to protect wires from abrasion and other types of damage. Wire mesh and other materials can also provide additional protection against abrasion to cables and pipelines. **Figure 10.3** illustrates corrugated tubes as part of a typical automotive harness.

Figure 10.3 Harness with corrugated tubes.

DreamStockP/Shutterstock.com.

- Protections against solid particles and dust: Filters and screens help prevent dust and other particles from entering sensitive components. Grilles and shields are also used to protect against larger debris.
- Thermal protection: Shields protect sensitive components from heat generated by the engine, exhaust system, and turbocharger. **Figure 10.4** illustrates an aluminum shield over a catalytic converter. Also noteworthy is the firewall between the engine compartment and the cabin.

Figure 10.4 Catalytic converter shield.

Ulianenko Dmitrii/Shutterstock.com.

- Protection against liquids: boxes, lids, covers, sealing joints, gutters/pipes/drainage ducts, shields, resins, varnishes, breathing pads, high installation point, and prediction/routing of drips. **Figure 10.5** exhibits a typical fuse and relay box in the engine compartment, with a protection lid.

10.7.
Electrical Barriers

Electrical barriers serve multiple purposes in vehicle design. They are used to prevent the spread of damage in case individual components and parts fail. These barriers are designed to isolate or contain electrical faults to stop them from affecting other parts of the vehicle's electrical system. Additionally, electrical barriers play a crucial role in limiting electrical stresses within the system. The following sections discuss some common electrical barriers and their purposes.

10.7.1.
Fuses

Traditional fuses are protective devices designed to interrupt the flow of electrical current when a circuit experiences an overcurrent or a short circuit. They are typically rated for a specific current and are placed in line with the circuit they protect. When a fault occurs, the fuse

Figure 10.5 Fuse box lid.

KT studio/Shutterstock.com.

"blows" or melts, breaking the circuit and preventing further electrical flow. This helps prevent damage to other components and systems connected to the same circuit [10.26].

10.7.2.
Resettable Fuses

Also known as polymeric positive temperature coefficient (PPTC) devices or polyswitches, resettable fuses are used for overcurrent and short circuit protection. In normal conditions, they present low resistance, allowing the operation of the loads. When a current surge or short circuit appears, the rise of the internal temperature rapidly makes them exhibit an increase in their series resistance, limiting the power supplied. After clearing the fault condition, they return to the low-resistance state. Some examples are seen in **Figure 10.6**.

Figure 10.6 Polyswitch fuses.

YouraPechkin/Shutterstock.com.

10.7.3.
Relays

Relays are electromagnetic switches that control the flow of current between two circuits, for example, the horn energizing after the depression of a low-current switch on the steering wheel. When directed by an electronic circuit, in case of a fault in the high-current circuit, the module can be designed to open the relay, preventing the fault from damaging the electric system more significantly. The black and gray cubes in **Figure 10.7** are automotive relays.

Figure 10.7 Relay and fuse box.

Yurii Kukharuk/Shutterstock.com.

10.7.4.
Bimetallic Thermal Switches

These thermal protectors prevent electric motors (e.g., window lifter motors) from overheating. They operate by utilizing the differing thermal expansion properties of two different metals. The switch is made up of two metal strips that are bonded together. The metals expand or contract at different rates when the temperature changes, causing the bimetal strip to bend. This bending action can be used to open the feed line of the motor, preventing its burn. As the temperature decreases, the bimetal strip will reverse its bending and close the circuit, and the motor can return to regular operation.

10.7.5.
Smart Switches

Several electronic components can be used to turn on and off a load, like bipolar transistors, field effect transistors, and thyristors. As the technology evolved, many of them are part of an integrated circuit, capable of detecting dangerous conditions (excessive current and excessive temperature, for instance) and disconnecting the load before permanent damage occurs. Smart switches are usually part of more complex electronic modules, commanding several loads.

10.7.6.
Surge Suppressors

Surge suppressors or transient voltage suppressors are used to protect sensitive electronic components from voltage spikes or transients. They divert excess voltage away from sensitive parts of the electrical system, preventing damage in the event of voltage surges, such as those caused by inductive loads or lightning. As technology evolved, several electronic parts fit into this category. Some

examples are Zener diodes, avalanche diodes, spark gaps, and metal oxide varistors.

10.7.7.
Electronic Architecture

In contemporary automotive electronic architectures, a key strategy involves selectively disconnecting loads and modules that are not essential when the vehicle is idling. This approach serves multiple crucial purposes. First, it significantly diminishes the current drawn from the battery, thus extending the period a car can remain parked without depleting its battery reserves. Additionally, this deliberate disconnection serves as an effective means of reducing the risk of fire.

The rationale behind this approach is grounded in reducing the probability of system failures. When electronic modules are deprived of power, their susceptibility to failures notably diminishes. Hence, this proactive measure conserves energy and enhances the vehicle's overall safety.

Furthermore, it is essential to consider the broader context: an unattended fire poses far more significant risks than when a driver is present. In the latter scenario, the driver can immediately address and mitigate the situation, potentially averting more extensive damage.

10.8.
Logical (Software) Barriers for xEVs

In software-equipped electronic modules, it is commonplace and cost-effective to implement monitoring, diagnosing, and protection features through software algorithms. In the specific case of xEVs, this gained even more prominence due to the increased complexity of their systems.

Some examples are discussed in the following sections.

10.8.1.
Battery System Management

Through sophisticated software algorithms, sophisticated control over battery temperature, voltage, charge, and discharge currents is achieved. This approach is extended, at least partially, to managing individual modules and cells within the battery.

10.8.2.
Smart Chargers

The deployment of intelligent chargers capable of dynamically adjusting charge and voltage levels to align with the specific vehicle, battery type, state of charge, and temperature ensures safe and efficient charging.

10.8.3.
Conductive Charger Immobilization

Implementing measures to immobilize the vehicle while connected to a conductive charger enhances user safety.

10.8.4.
Secure Charger Connection and Disconnection

Prioritizing operators' safety by designing systems that facilitate secure and arcing-minimized charger connection and disconnection. As implied by Figure 10.8, the safe connection and disconnection of a charger plug emerge as a pivotal aspect of the safety framework.

Figure 10.8 Woman handling a charging plug with a girl.

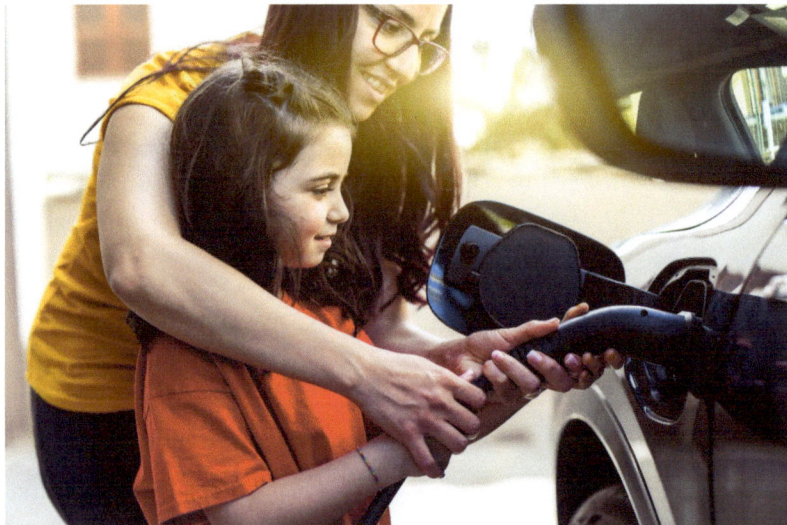

William Perugini/Shutterstock.com.

10.8.5.
Resilience to Failures

The implementation of failure tolerance mechanisms and mitigation measures is essential to enhance system reliability and minimize the impact of unexpected malfunctions.

10.8.6.
Emergency Battery Disconnection

The automatic disconnection of the high-voltage battery in the event of collisions or electronic module failures helps mitigate potential safety hazards.

10.9.
Traction Battery Design Layers

Lithium batteries featuring liquid electrolytes are prominent in today's landscape of mass-produced traction batteries for xEVs. Regrettably, this particular component type has been associated with several fire incidents, underscoring the need to enhance safety across its design, production, and application.

As elucidated by Larsson et al., achieving robust battery safety entails implementing a multi-faceted approach comprising several layers of safety techniques. **Figure 10.9** visually illustrates this safety onion, with a hierarchy of the diverse measures deployed to reduce the likelihood of faults and mitigate the potential consequences arising from such faults [10.27, 10.28].

Figure 10.9 Safety layers of xEV battery systems.

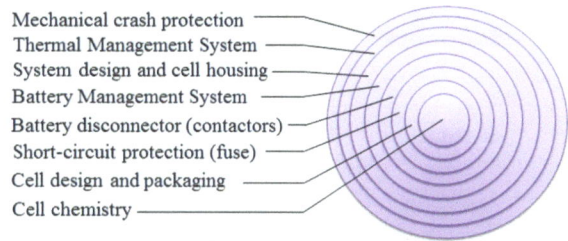

Larsson, Fredrik & Andersson, Petra & Mellander, B.-E. (2016). Lithium-Ion Battery Aspects on Fires in Electrified Vehicles on the Basis of Experimental Abuse Tests. Batteries. 2. 9. 10.3390/batteries2020009. Licensed under CC BY 4.0 https://creativecommons.org/licenses/by/4.0/.

References

[10.1].   Gerwin, D. and Barrowman, N., "An Evaluation of Research on Integrated Product Development," *Management Science* 48, no. 7 (2002): 938-953, doi:https://doi.org/10.1287/mnsc.48.7.938.2818.

[10.2].   Tang, H., *Automotive Vehicle Assembly Processes and Operations Management*, SAE International Book R-456 (Warrendale: SAE International, 2017), ISBN:978-0-7680-8339-2.

[10.3].   Womack, J., Jones, D., and Roos, D., *The Machine That Changed the World* (London, UK: Simon & Schuster, 2007), ISBN:978-0743299794.

[10.4].   Jaswal, A., Rajasekhar, M., Perumal, J., and Rawte, S., "Aspects of Fire and Thermal Safety in Vehicle Development," SAE Technical Paper 2015-26-0156 (2015), doi:https://doi.org/10.4271/2015-26-0156.

[10.5].   SAE International Ground Vehicle Standard, "Guidelines for Electric Vehicle Safety," SAE Standard J2344, October 2020.

[10.6].   International Organization for Standardization, "Electrically Propelled Road Vehicles—Safety Specifications—Part 1: Rechargeable Energy Storage System (RESS)," Standard ISO 6469-1, 2019.

[10.7]. International Organization for Standardization, "Electrically Propelled Road Vehicles—Safety Specifications—Part 2: Vehicle Operational Safety," Standard ISO 6469-2, 2022.

[10.8]. International Organization for Standardization, "Electrically Propelled Road Vehicles—Safety Specifications—Part 3: Electrical Safety," Standard ISO 6469-3, 2021.

[10.9]. International Organization for Standardization, "Electrically Propelled Road Vehicles—Safety Specifications—Part 4: Post-Crash Electrical Safety," Standard ISO 6469-4, 2015.

[10.10]. International Organization for Standardization, "Road Vehicles—Test Methods for Electrical Disturbances from Electrostatic Discharge," Standard ISO 10605, 2023.

[10.11]. International Organization for Standardization, "Road Vehicles—Environmental Conditions and Testing for Electrical and Electronic Equipment," Standard ISO 16750, 2018.

[10.12]. SAE International Ground Vehicle Standard, "Automobile and Motor Coach Wiring," SAE Standard J1292, April 2016.

[10.13]. SAE International Ground Vehicle Standard, "High Voltage Automotive Wiring Assembly Design," SAE Standard J1673, March 2012.

[10.14]. SAE International Ground Vehicle Standard, "Electric-Drive Battery Pack System: Functional Guidelines," SAE Standard J2289, August 2021.

[10.15]. SAE International Ground Vehicle Standard, "Safety Standard for Electric and Hybrid Vehicle Propulsion Battery Systems Utilizing Lithium-based Rechargeable Cells," SAE Standard J2929, February 2013.

[10.16]. National Highway Traffic Safety Administration (NHTSA), "Fuel System Integrity," FMVSS 301, Code of Federal Regulations, Federal Motor Vehicle Safety Standards, 49 CFR571.301 Standard No. 301, April 2004.

[10.17]. National Highway Traffic Safety Administration (NHTSA), "Flammability of Interior Materials," FMVSS 302, Code of Federal Regulations, Federal Motor Vehicle Safety Standards, 49 CFR 571.302 Standard No. 302, October 2021.

[10.18]. National Highway Traffic Safety Administration (NHTSA), "Fuel System Integrity of Compressed Natural Gas Vehicles," FMVSS 303, Code of Federal Regulations, Federal Motor Vehicle Safety Standards, 49 CFR § 571.303—Standard No. 303, November 1995.

[10.19]. National Highway Traffic Safety Administration (NHTSA), "Compressed Natural Gas Fuel Container Integrity," FMVSS 304, Code of Federal Regulations, Federal Motor Vehicle Safety Standards, 49 CFR § 571.304—Standard No. 304, February 2022.

[10.20]. National Highway Traffic Safety Administration (NHTSA), "Electric-Powered Vehicles, Electrolyte Spillage and Electrical Shock Protection," FMVSS 305, Code of Federal Regulations, Federal Motor Vehicle Safety Standards, 49 CFR 571.305—Standard No. 305, August 2019.

[10.21]. European Automobile Manufacturers' Association (ACEA), "Regulatory Guide," accessed October 2023, https://www.acea.auto/files/ACEA-Regulatory-Guide-2023.pdf.

[10.22]. United Nations Economic Commission for Europe (UNECE), "Prevention of Fire Risks (Liquid Fuel Tanks)," European Community Regulation EC No. 2019/2144, UN R34, 2019.

[10.23]. United Nations Economic Commission for Europe (UNECE), "Battery Electric Vehicles Safety," European Community Regulation EC No. 2019/2144, UN R100, 2019.

[10.24]. United Nations Economic Commission for Europe (UNECE), "Compressed Natural Gas (CNG) Systems," European Community Regulation EC No. 2019/2144, UN R110, 2019.

[10.25]. United Nations Economic Commission for Europe (UNECE), "Fire Resistance of Interior Materials," European Community Regulation EC No. 2019/2144, UN R118, 2019.

[10.26]. SAE International Ground Vehicle Standard, "Blade Type Electric Fuses," SAE Standard J1284, April 1988.

[10.27]. Chang, C., Gorin, C., Zhu, B., Beaucarne, G. et al., "Lithium-Ion Battery Thermal Event and Protection: A Review," *SAE Int. J. Elect. Veh.* 13, no. 3 (2024): 1-41, doi:https://doi.org/10.4271/14-13-03-0019.

[10.28]. Larsson, F., Andersson, P., and Bengt-Erik, M., "Battery Aspects on Fires in Electrified Vehicles," in *Proceedings from 3rd International Conference on Fires in Vehicles—FIVE 2014*, Berlin, Germany, January 1, 2014, 209-220, https://publications.lib.chalmers.se/publication/204919-battery-aspects-on-fires-in-electrified-vehicles.

Chapter 11

Development and Maintenance

11.1.
Introduction

In the ever-evolving world of automotive engineering, developing safe, reliable, and innovative vehicles is a multifaceted challenge that requires a systematic and rigorous approach [11.1-11.7]. This chapter covers the V-cycle development process and recognizes that computer simulations and validations are the bedrock of modern automotive development activities, while physical testing remains an indispensable cornerstone. It examines the stages of progressive product and process development and acknowledges the significance of poka-yoke techniques, risk evaluation, and mitigation strategies. Finally, it delves into maintenance, an often overlooked yet undeniably imperative aspect.

While delivering an array of benefits, each of these processes and strategies collectively contributes to mitigating the risks entailed by car fires.

11.2.
V-Cycle

The "V-Cycle" refers to a systematic and structured approach to the product development process. It is also known as the "V-Model" or "Verification and Validation Model." The V-cycle is a widely used framework in various industries, including automotive, to ensure that a product—in this case, a vehicle—meets the required specifications and quality standards while minimizing errors and issues [11.8, 11.9].

The V-cycle is called so because of its visual representation, which looks like the letter "V." The model is typically divided into two main phases:

the left side of the "V" represents the development phase and the right side represents the validation and verification phase. The horizontal axis (usually not depicted) implies the passage of time. The following sections describe a breakdown of the key stages within each side of the "V."

11.2.1.
Left Side—Development Phase

11.2.1.1.
Requirements Definition

At the top of the "V," the product's requirements are formally established and documented. These requirements serve as the starting point for the entire development process.

11.2.1.2.
System Design

Once the requirements are defined, the system architecture and high-level design are created. This stage outlines how the product will meet the specified requirements.

11.2.1.3.
Component Design

The high-level design is further broken down into subsystems and individual components, and detailed designs are created for each of them.

11.2.1.4.
Implementation

This is where the actual development and manufacturing of the automotive systems and components occur based on the designs and specifications.

11.2.2. Right Side—Verification and Validation Phase

11.2.2.1.
Unit Testing

The individual components and subsystems are rigorously tested to ensure they function as intended.

11.2.2.2.
Integration Testing

Components are integrated and tested to ensure they work together as a complete system.

11.2.2.3.
System Testing

The entire vehicle system is tested to verify that it meets the original requirements and specifications.

11.2.2.4.
Validation Testing

This phase focuses on ensuring that the product meets the needs of the end users and complies with regulatory and safety standards. It involves testing in real-world conditions.

11.2.2.5.
Release and Maintenance

Once the product successfully passes all tests and validations, it is released for production. Maintenance and updates may continue throughout the product's life cycle.

The V-cycle emphasizes the importance of testing and validation activities in parallel with the development process. It ensures that potential issues are identified early in the development cycle and can be addressed before they become costly to fix. It also promotes traceability, as each stage on the left side of the "V" should have a corresponding validation or testing stage on the right side. **Figure 11.1** shows an example of the application of this process in powertrain development.

The V-cycle is essential in automotive design and development to improve product quality, reduce risks in general (including fire hazards), and ensure that vehicles meet safety, performance, and regulatory standards.

Figure 11.1 Example of the V-cycle in powertrain development.

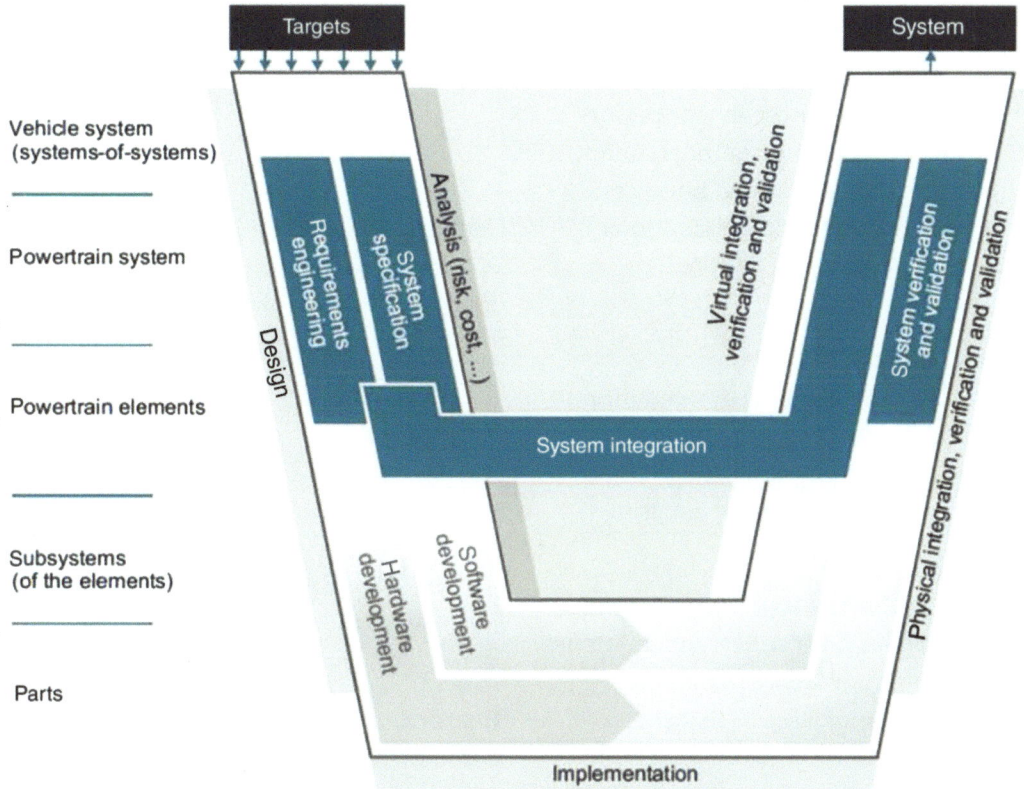

11.3.
Simulations

Computer simulations and virtual validations play a crucial role in the automotive development process for a variety of reasons [11.10-11.12].

11.3.1.
Cost-Efficiency
Traditional physical testing and prototyping are expensive, time-consuming, and resource-intensive. Computer simulations and virtual validations allow automotive manufacturers to reduce the need for physical prototypes, saving time and money.

11.3.2.
Rapid Prototyping
Virtual simulations enable the rapid creation and testing of prototypes. This agility allows automotive companies to iterate and refine designs more quickly.

11.3.3.
Safety Testing
Virtual simulations are used to test vehicle safety in various scenarios, such as crash tests and collision simulations. These tests can identify

potential safety issues and allow for improvements before physical testing.

11.3.4.
Performance Optimization

Computer simulations optimize vehicle performance, including fuel efficiency, aerodynamics, temperature profiles, and handling. Engineers can experiment with different configurations and select the best-performing options.

11.3.5.
Environmental Impact

Virtual validations can help reduce the environmental impact of vehicle development by minimizing the need for physical prototypes and tests, which can generate waste and emissions.

11.3.6.
Regulatory Compliance

Automotive companies must adhere to numerous safety and emissions regulations. Computer simulations enable manufacturers to test and ensure compliance with these standards more efficiently.

11.3.7.
Early Issue Identification

Simulations can reveal design issues and shortcomings early in the development process, making it easier and less expensive to address problems before they become critical.

11.3.8.
Noise and Vibration Analysis

Virtual validations are used to analyze and minimize noise and vibration issues, improving vehicle comfort and quality and reducing vibration stresses on components.

11.3.9.
Testing Extreme Conditions

Simulations can model extreme conditions that are difficult or dangerous to replicate physically, such as extreme temperatures or off-road environments.

11.3.10.
Design Verification

Virtual validations can verify that the vehicle's design will meet its intended purpose and function as expected. This includes the evaluation of complex systems like autonomous driving technologies.

11.3.11.
Human–Machine Interface (HMI)

Simulations are used to test vehicles' HMIs, ensuring they are intuitive, user-friendly, and safe.

11.3.12.
Training and Education

Automotive manufacturers use simulations for employee training and skill development. This allows employees to become proficient with new technologies and processes before they are implemented in real-world production.

Figure 11.2 illustrates the use of 3D-enhanced reality in developing a brake system.

Overall, computer simulations and virtual validations are essential tools for automotive development. They offer efficiency, cost savings, and the ability to create safer, more reliable, and better-performing vehicles. These technologies are becoming increasingly sophisticated, enabling engineers to model and validate complex automotive systems with greater accuracy and confidence.

Figure 11.2 Enhanced reality example.

Gorodenkoff/Shutterstock.com.

11.4.
Physical Validations

Physical validations are still employed in the automotive development process, and the industry does not rely solely on computer simulations for several vital reasons [11.13-11.16], as discussed in the following sections.

11.4.1.
Real-World Verification

Physical validations involve building and testing actual prototypes or components of a vehicle. These tests provide real-world verification of how they perform in physical conditions. This is vital because it confirms that the product functions as intended and meets safety, performance, and regulatory requirements.

11.4.2.
Unknown and Unpredictable Variables

While computer simulations are powerful tools for modeling and predicting various scenarios, they are curtailed by the accuracy of the data and the complexity of the physical world. Many real-world variables, such as road conditions, weather, and human behavior, are challenging to simulate accurately. Physical tests account for these unpredictable factors.

11.4.3.
Complex Interactions

Automotive systems are complex and interconnected. Computer simulations can model individual components and their interactions. Still, the modeling process often has unforeseen interactions and inaccuracies that can only be discovered through physical validations.

11.4.4.
Safety Assurance

Safety is paramount in the automotive industry. Physical validations help ensure a vehicle meets safety standards and can withstand crash tests and other assessments. Human lives are at stake, so rigorous physical testing is essential.

11.4.5.
Quality Assurance

Physical validations can uncover manufacturing defects, material weaknesses, or other issues that might not be apparent in simulations. Ensuring the quality of each component and the vehicle as a whole is a fundamental requirement.

11.4.6.
Validation of Prototypes

Physical prototypes and mock-ups are often tested for ergonomics, user experience, aesthetics, and customer abuse. These aspects are challenging to assess accurately through simulations.

11.4.7.
Regulatory Compliance

Automotive products must adhere to strict regulations and standards. Physical tests are often demanded to demonstrate compliance with these standards. **Figure 11.3** shows a frontal crash test example.

Figure 11.3 Frontal crash test.

Benoist/Shutterstock.com.

11.4.8.
Customer Satisfaction

Real-world testing helps ensure a vehicle meets customer expectations and requirements. Factors like ride comfort, noise levels, and overall user experience are best evaluated in physical validations.

11.4.9.
Benchmarking and Competitive Analysis

Physical testing is used for benchmarking vehicles against competitors before and after commercial launch. Manufacturers must understand how their products compare performance, durability, and other objective and subjective factors in the real world.

11.4.10.
Iterative Development

The automotive development process often involves iterative design and testing cycles. Physical validations allow for feedback and adjustments to design and engineering, eventually saving time and resources compared to repeatedly revising computer modeling and simulations.

While computer simulations are valuable tools for early-stage design and concept validation, physical validations are essential for the final confirmation of a product's suitability for the real world. The automotive industry employs both methods to balance cost-efficiency and thorough testing, ensuring their vehicles' safety, quality, and performance, all under a given tight timeline of the project.

11.5.
Product and Process Development Stages

Product and process development stages are critically important in the automotive industry to ensure the successful design, development, and production of vehicles. These stages involve a series of steps, each with its specific objectives and prototypes, and they play a vital role in bringing a new car from concept to mass production. The following sections provide an explanation of their relevance and the use of progressively representative prototypes and processes, as well as the customization of these stages by automotive manufacturers.

11.5.1.
Relevance of Product and Process Development Stages

11.5.1.1.
Product Development Stages

These stages focus on designing and engineering the vehicle. They are essential for ensuring the car meets customer requirements, safety standards, and performance expectations.

11.5.1.2.
Process Development Stages

These stages focus on developing the manufacturing processes, assembly lines, and quality control procedures necessary to produce the vehicle at scale. Process development is critical for achieving efficiency, cost-effectiveness, and consistency in manufacturing.

11.5.1.3.
Integration of Product and Process Development

Product and process development are closely intertwined. The product design should align with the manufacturing processes, and any changes made to the product (e.g., design modifications) should be reflected in the manufacturing processes.

11.5.2.
Progressively Representative Prototypes and Processes

11.5.2.1.
Mules

Mules are early-stage prototypes used in the automotive industry. They are typically hand-built vehicles to test and validate critical components and systems. Frequently, they are constructed by adapting similar parts from existing, mass-produced vehicles. Mules are not production-ready but help evaluate the feasibility of the vehicle's basic design and functionality.

11.5.2.2.
Pre-Series Prototypes
Pre-series prototypes are closer to the final production version of the vehicle. They are used to validate the design, manufacturing processes, and quality control procedures. Pre-series prototypes often include specialized tooling and equipment used in their production process.

11.5.2.3.
Mass Production Prototypes
These prototypes are the closest representation of the final production vehicle. They are used for validation of the entire manufacturing process, including the assembly line and quality control. Mass production prototypes are subjected to extensive testing to ensure that the product and process meet all specifications and standards, with the production process running at the maximum rate.

11.5.3.
Customization of Development Stages

11.5.3.1.
Each Automotive Manufacturer's Approach
Automotive manufacturers often customize their product and process development stages based on their specific needs, product lines, and development philosophies. While the core stages are consistent, the details and the number of intermediate phases may vary.

11.5.3.2.
Intermediate Phases
Some manufacturers may introduce additional stages or phases, such as alpha and beta prototype phases, which are intermediate between mules and pre-series. The number of

phases also depends on the overall project scope, that is, a simple aesthetic model year evolution requires fewer stages, while a brand-new vehicle model usually demands more phases. These other phases allow for more extensive testing, validation, and refinements before more significant tooling and manufacturing investments are made.

11.5.3.3.
Iterative Approach
The automotive development process is often iterative, with feedback from testing and validation leading to design and process refinements. This iterative approach helps improve the final product's quality and reliability.

The product and process development stages of the automotive industry are integral to producing safe, reliable, and market-ready vehicles. The progressive representation of prototypes and processes ensures that issues are identified and resolved at an early stage, reducing the likelihood of costly design changes after mass production has begun. Each automotive manufacturer customizes these stages to suit specific requirements and production methods, often incorporating their intermediate phases and quality control checkpoints.

11.6.
Poka-Yoke

Poka-yokes, also known as mistake-proofing or error-proofing techniques, can be exemplified by using different incompatible connectors, as seen in **Figure 11.4**. These techniques have significant relevance in the current automotive industry for several reasons [11.17].

Figure 11.4 Different USB connectors.

Nopparat S/Shutterstock.com.

11.6.1.
Quality Assurance

Poka-yokes play a crucial role in ensuring product quality. In the automotive industry, quality is paramount to meet safety standards, customer expectations, and regulatory requirements. Poka-yokes help prevent defects and errors at various stages of the production process, enhancing the overall quality of vehicles.

11.6.2.
Cost Reduction

Automobile manufacturing mistakes can be costly, especially if detected after assembling the vehicle. Poka-yokes help in the early detection and correction of errors, reducing the need for rework or expensive recalls. This results in significant cost savings for automotive manufacturers.

11.6.3.
Efficiency and Productivity

Poka-yokes streamline production processes by eliminating the need for manual inspection and correction of defects. This increases efficiency and productivity as workers can focus on value-added tasks, rather than error detection and correction.

11.6.4.
Worker Safety

Many poka-yoke mechanisms are designed to enhance worker safety. By preventing mistakes that could lead to accidents or injuries on the assembly line, poka-yokes contribute to a safer work environment in the automotive industry.

11.6.5.
Regulatory Compliance

The automotive industry is subject to strict safety and environmental regulations. Mistakes or defects in vehicle components can lead to regulatory noncompliance, resulting in fines and damage to a manufacturer's reputation. Poka-yokes help ensure compliance with these regulations by reducing the likelihood of errors that could lead to noncompliance.

11.6.6.
Customer Satisfaction

High-quality vehicles with fewer defects lead to increased customer satisfaction. Poka-yokes help manufacturers consistently produce cars that meet or exceed customer expectations, leading to better brand reputation and customer loyalty.

11.6.7.
Complexity of Modern Vehicles

In the contemporary automotive landscape, vehicles have achieved unprecedented levels of sophistication, seamlessly incorporating advanced systems, electronics, sensors, and intricate safety features. The inherent intricacy of these cutting-edge technologies naturally amplifies the potential for errors during assembly and integration processes. Recognizing this challenge, adopting Poka-yokes is an indispensable practice in modern automotive manufacturing, effectively mitigating these risks.

These error prevention mechanisms are pivotal in eliminating or substantially reducing the likelihood of incorrect assembly operations. In doing so, Poka-yokes significantly contribute to the enhancement of overall manufacturing precision and play a key role in ensuring the reliability of the end product.

11.6.8.
Lean Manufacturing

Poka-yokes are a vital component of lean manufacturing principles, which aim to eliminate waste and improve efficiency. By preventing errors and defects, they contribute to lean practices widely adopted in the automotive industry.

In summary, poka-yokes are highly relevant in the current automotive industry as they contribute to improved quality, cost reduction, efficiency, safety, and compliance with regulations. They are essential in maintaining high standards in vehicle production and ensuring customer and regulatory bodies' satisfaction.

11.7.
Risk Evaluations

Risk evaluation is of paramount importance throughout the automotive development process. Automotive development involves complex systems, integration of numerous components, definition and implementation of several manufacturing processes, and compliance with stringent safety and performance standards. Therefore, identifying, assessing, and managing risks are crucial to ensure the successful and safe deployment of automotive products. The following sections discuss some key reasons why risk evaluations are relevant in automotive development.

11.7.1.
Safety Assurance

Safety is the paramount concern in the automotive industry, encompassing both manufacturing and end-user perspectives. In the manufacturing environment, safety risks extend to the operator's well-being, requiring evaluating potential hazards associated with manual assembly processes, machinery operation, and ergonomic considerations. Identifying and mitigating these risks ensure a safe working environment for operators, contributing to overall safety assurance in the automotive manufacturing process. On the end-user front, risks such as design flaws, component failures, or software errors can lead to accidents and life-threatening situations, including fire hazards. Risk evaluations are crucial in preemptively identifying and mitigating these potential safety hazards, underscoring their significance in fostering a secure and reliable automotive environment for manufacturers and end customers.

11.7.2.
Compliance with Regulations

The automotive industry is heavily regulated, with numerous safety, environmental, and quality standards to meet. Risk assessments help ensure a product complies with these regulations, reducing the likelihood of costly legal consequences or product recalls.

11.7.3.
Complex Systems

Modern vehicles are highly complex systems incorporating numerous subsystems, electronic components, and software. Risk evaluations help identify and address potential integration issues and interdependencies that could compromise the vehicle's functionality.

11.7.4.
Factory Floor Evaluations

Conducting thorough evaluations on the factory floor is essential in developing logistic and manufacturing processes. This involves assessing risks associated with water intrusion, contamination (including dust or other substances), and potential damage due to mishandling—from the warehouse to the assembly line installation point, encompassing intermediate stocks and subassemblies. Failing to address these concerns may lead to the accidental installation of damaged or contaminated parts in vehicles, presenting safety and fire risks once delivered. This underscores the importance of meticulous evaluations to maintain the integrity and safety of automotive components throughout the manufacturing journey.

11.7.5.
Cost Control

Identifying and managing risks early in the development process can help control costs. Addressing issues in the later stages of development or after a product is in the market can be extremely costly. Risk assessments enable cost-effective problem-solving and resource allocation.

11.7.6.
Quality Assurance

Risks can have a detrimental impact on the quality of the final product. Identifying risks and mitigating them early in the development process ensure that the end product meets quality standards and customer expectations.

11.7.7.
Market Competitiveness

Delivering a reliable and safe product in the highly competitive automotive market is essential to gain a competitive edge. Risk evaluations help ensure a vehicle can perform reliably under various conditions and meet customer expectations.

11.7.8.
Reputation and Brand Trust

Automotive companies heavily rely on their reputation and brand trust. Vehicle recalls or safety issues can significantly damage a brand's image. Proactive risk management safeguards a company's reputation and confidence in the market.

11.7.9.
Supplier Management

Many automotive components come from suppliers. Evaluating the risks associated with the supply chain, including supplier reliability and quality, is vital to prevent disruptions and maintain consistency in the manufacturing process.

11.7.10.
Environmental Impact

Risk assessments can also consider environmental factors, helping automotive companies address potential environmental concerns related to vehicle production, emissions, and disposal.

In summary, risk evaluations are integral to ensuring the success of automotive projects by identifying potential issues early in the process, ensuring compliance with safety and quality standards, controlling costs, managing project schedules, and safeguarding the company's reputation. By proactively addressing risks, automotive manufacturers can create safer, more reliable, and competitive products while minimizing potential financial and legal liabilities.

11.8.
Maintenance

Proper maintenance is crucial in the automotive context to avoid the risk of fires and ensure the safety and reliability of vehicles. This includes regular inspections, preventive maintenance, and corrective maintenance. Even if a car is perfectly designed and produced, neglecting maintenance can lead to increased risks associated with degradation, wear and tear, accidents, misuse, and other factors [11.18]. The following sections provide an explanation of the relevance of proper maintenance in the automotive context.

11.8.1.
Regular Inspections

Regular vehicle inspections are essential to identify potential issues before they escalate into serious problems. Regarding fire avoidance, inspections can uncover issues like oil or fuel leaks, damaged wiring, or overheating components. Detecting these issues early can prevent them from becoming fire hazards. Additionally, inspections can identify worn-out brake pads, tires, or other safety-critical components that can lead to accidents.

11.8.2.
Preventive Maintenance

Preventive maintenance involves proactive measures to prevent issues from occurring. This can include changing the oil and oil filters, replacing spark plugs, checking and replacing worn-out belts and hoses, inspecting the cooling system, and removing solid particles around the exhaust system (especially in agricultural machinery). Preventive maintenance helps keep the vehicle in optimal conditions, reducing the risk of components failing or malfunctioning, which can lead to fires or accidents.

11.8.3.
Corrective Maintenance

Corrective maintenance is performed when issues are identified during inspections or when a component has failed. Addressing these issues promptly is essential for safety. For example, if a vehicle's electrical system is not working correctly, it can result in short circuits or overheating, potentially leading to a fire. Proper corrective maintenance ensures that faulty components are repaired or replaced to prevent such incidents.

11.8.4.
Risks Associated with Neglect

Neglecting maintenance can lead to increased risks of fires, accidents, and other safety hazards. For instance, an ignored oil leak can result in oil buildup, which is highly flammable. Hot gases escaping from a damaged exhaust system can lead to fires. Worn-out brake components can result in poor braking performance, increasing the risk of accidents. Ignoring any of these risks can have severe consequences.

11.8.5.
Electronic Diagnostics

Modern vehicles have complex electronic systems; electronic diagnostics are crucial for identifying and addressing issues. These systems can detect problems with engine performance, wiring issues, and safety features. Regularly connecting a vehicle to diagnostic equipment, as exemplified in **Figure 11.5**, can help identify and resolve problems that may not be apparent during visual inspections.

In summary, proper maintenance is essential for fire avoidance, as it helps detect and address issues that can lead to fires and accidents. Even with a perfectly designed and produced vehicle, neglecting maintenance can introduce

Figure 11.5 Technician running electronic diagnostics.

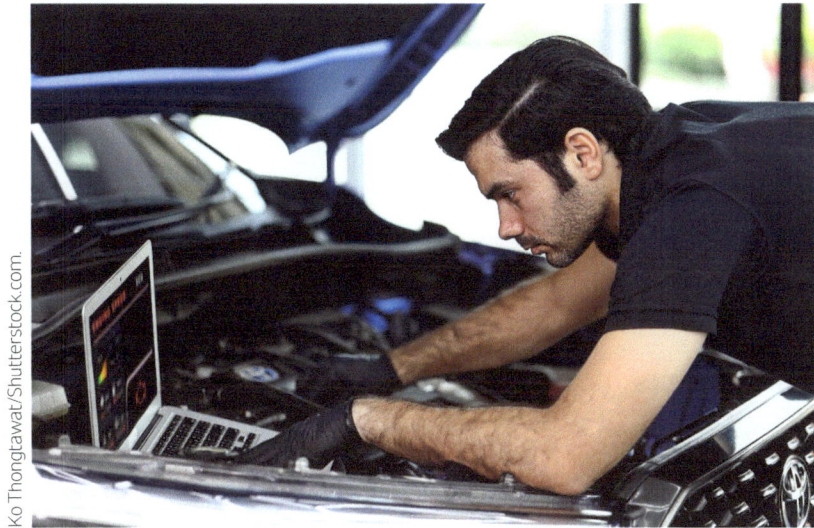

significant risks. Regular inspections, preventive maintenance, and corrective maintenance are all essential components of a comprehensive maintenance program that ensures the safety and reliability of vehicles. Additionally, electronic diagnostics play a vital role in identifying hidden issues and maintaining the overall health of modern cars.

References

[11.1].   Ball, J. and Stone, R., *Automotive Engineering Fundamentals*, SAE International R-199 Book (Warrendale: SAE International, 2004), ISBN:978-0-7680-0987-3.

[11.2].   El-Sayed, M., *Fundamentals of Integrated Vehicle Realization*, SAE International R-436 Book (Warrendale: SAE International, 2017), ISBN:978-0-7680-8383-5.

[11.3].   Jaswal, A., Rajasekhar, M., Perumal, J., and Rawte, S., "Aspects of Fire and Thermal Safety in Vehicle Development," SAE Technical Paper 2015-26-0156 (2015), doi:https://doi.org/10.4271/2015-26-0156.

[11.4].   Kossiakoff, A., Sweet, W., Seymour, S., and Biemer, S., *Systems Engineering Principles and Practice*, 2nd ed. (New York: John Wiley & Sons, 2011), doi:https://doi.org/10.1002/9781118001028.

[11.5].   SAE International Ground Vehicle Standard, "Safety Standard for Electric and Hybrid Vehicle Propulsion Battery Systems Utilizing Lithium-Based Rechargeable Cells," SAE Standard J2929, February 2013.

[11.6].   SAE International Ground Vehicle Standard, "Hybrid and Electric Vehicle Safety Systems Information Report," SAE Standard J2990/2, J2990/2_202011, November 2020.

[11.7].   Seiffert, U. and Gonter, M., *Integrated Automotive Safety Handbook*, SAE International R-407 Book (Warrendale: SAE International, 2013), ISBN:978-0-7680-8029-2.

[11.8].   Leighton, M. and Ray, R., "Verification and Validation for Modular Development Platforms," SAE Technical Paper 2023-01-0476 (2023), doi:https://doi.org/10.4271/2023-01-0476.

[11.9].   SAE International Ground Vehicle Standard, "Handbook for Robustness Validation of Automotive Electrical/Electronic Modules," SAE Standard J1211, November 2012.

[11.10]. Gu, L. and Yang, R., "CAE Model Validation in Vehicle Safety Design," SAE Technical Paper 2004-01-0455 (2004), doi:https://doi.org/10.4271/2004-01-0455.

[11.11]. Aruna Devi, D., Veeramanikandan, M., Vavilapalli, K.R., and Jeevan, N.K., "HiL Testing of Automotive Transients in Electric Vehicle," SAE Technical Paper 2022-28-0378 (2022), doi:https://doi.org/10.4271/2022-28-0378.

[11.12]. Joshi, A., *Automotive Applications of Hardware-in-the-Loop (HIL) Simulation*, SAE International PT-209 Book (Warrendale: SAE International, 2020), ISBN:978-1-4686-0003-2.

[11.13]. SAE International Ground Vehicle Standard, "Recommended Practice for Electric, Fuel Cell and Hybrid Electric Vehicle Crash Integrity Testing," SAE Standard J1766, January 2014.

[11.14]. SAE International Ground Vehicle Standard, "Recommended Testing Methods for Physical Protection of Wiring Harnesses," SAE Standard J2192, June 2021.

[11.15]. SAE International Ground Vehicle Standard, "Vibration Testing of Electric Vehicle Batteries," SAE Standard J2380, December 2021.

[11.16]. SAE International Ground Vehicle Standard, "Electric and Hybrid Electric Vehicle Rechargeable Energy Storage System (RESS) Safety and Abuse Testing," SAE Standard J2464, August 2021.

[11.17]. Productivity Press Development Team, *Mistake-Proofing for Operators: The ZQC System* (Portland, OR: Productivity Press, 1997), ISBN:9781563271274.

[11.18]. Gilles, T., *Automotive Service Inspection, Maintenance, Repair*, 6th ed. (Boston: Cengage Learning, 2020), ISBN13: 978-1-337-79403-9.

Section 3: Analysis

Chapter 12

Recommended Analysis Method

12.1.
Introduction

Before delving into car fire analysis, readers are encouraged to explore local entities that offer specific training and certification tailored to their respective countries. Shifting to the focus of this chapter, one should consider that understanding the overarching goal of fire analysis and identifying the root cause or causal nexus are paramount. Root cause analysis in the automotive industry systematically pinpoints issues, while forensic engineering employs "causal nexus" to denote the intrinsic relationship between events and consequences.

The scientific method, a logical approach, is indispensable in investigating car fires, ensuring precision, evidence-based reasoning, and collaboration. A taxonomy of fire categories, from accidental to intentional, is outlined, emphasizing diverse contributing factors. This chapter details each step of the analysis, highlighting the importance of the systematic investigation. Cautionary notes address biases, stressing unbiased data collection and rigorous testing. A staged approach is recommended to mitigate risks, starting with identifying the starting location of the fire, determining the root cause or causal nexus, and, if necessary, discerning accidental or intentional events.

Exceptions, like emergencies, are acknowledged, but the chapter underscores applying scientific principles where possible.

In certain countries, particular care must be taken regarding the chain of evidence, also known as the chain of custody. Ultimately, the quest for the cause demands a judicious blend of scientific rigor and practical considerations.

12.2.
Fire-Related Institutions

Numerous institutions worldwide are dedicated to advancing fire knowledge, providing resources, establishing standards, offering training, overseeing control measures, and certifying investigators [12.1]. In the dynamic landscape of fire-related information and evolving technologies, it is crucial to regularly verify the relevance of available resources specific to your geographical location. Below is a concise list of accredited entities offering diverse expertise and support in the field.

- Asian Institute of Fire Safety, India.
- Bureau of Fire Protection (BFP), Philippines.
- Canadian Association of Fire Investigators (CAFI), Canada.
- Confederation of Fire Protection Associations Europe (CFPA), Europe.
- Danish Institute of Fire and Security Technology (DBI), Denmark.
- European Forest Institute (EFI), in several European countries.
- EV Fire Safe, Australia.
- International Association of Arson Investigators (IAAI), headquartered in the US.
- International Association of Fire & Rescue Services (CTIF, former *Comité Technique*

International de prevention et d'extinction de Feu), headquartered in Slovenia.

- International Fire Chiefs Association (IAFC), headquartered in the US.
- International Association of Fire Investigators (IAFI), headquartered in the US.
- International Fire Safety Standards Coalition (IFSSC), headquartered in Switzerland.
- National Association of Fire Investigators (NAFI), headquartered in the US.
- National Fire Protection Association (NFPA), US.
- National Forensic Science Training Institute (NFSTI), Philippines.

12.3.
The Search for the Cause

The typical goal of the analysis process is to ascertain the incident's root cause (utilizing terminology commonly employed in the automotive industry) or the causal nexus (by the terminology commonly used in forensic engineering and legal contexts). Regardless of whether the incident resulted in damage to a single vehicle component or the complete engulfing of an entire vehicle–carrier ship in flames, investigators are urged to set aside the impact of the damage magnitude. Instead, they should channel their efforts into accurately identifying the ignition point of the fire and its fundamental cause.

Root cause analysis is a systematic process that employs various problem-solving techniques to pinpoint the underlying source of an issue, such as equipment failure. This method typically seeks to provide insights into when, how, and why the problem initially arose, often leading to developing and implementing practical solutions to prevent its recurrence [12.2].

Conversely, in fields like forensic science, insurance inspection, and the legal system, the central focus is establishing the causal nexus. The term "causal nexus" holds legal significance, denoting the intrinsic relationship between an event and its subsequent consequences. In forensic science, investigators might leverage the causal nexus to establish a connection between a suspect and a crime scene or victim. At the same time, insurance inspectors utilize it to determine whether an insured event falls within the scope of coverage offered by an insurance policy. In the justice system context, the causal nexus plays a pivotal role in apportioning responsibility and ensuring fair adjudication of events.

While considering these fundamental concepts, it is essential to recognize that the definition of the root cause itself may exhibit variability, contingent on the specific taxonomy employed. To illustrate this point, the causative event can be straightforwardly categorized as "intentional" or "accidental," or more detailed categories could be employed. For the moment, let us enumerate some of the car fire categories utilized in the automotive industry and forensic analysis:

- Accidental.
- Arson (deliberate acts where an individual intentionally ignites a car). Note: Its definition varies from one jurisdiction to another.
- Braking system issues.
- Collision.
- Cooling system failures.
- Defective parts.
- Design errors.
- Electrical issues (encompassing various subcategories such as faulty connections and wiring, short circuits, overloads, and electronic module malfunctions).
- Exhaust system issues.
- External sources (involving nearby fires, lightning strikes, and other external factors).
- Faulty accessories (e.g., a malfunctioning cell phone charger).
- Flammable cargo.
- Fuel system leakages.
- Leakage of other combustible fluids.
- Improper accessory installation.
- Intentional.
- Maintenance errors.
- Maintenance neglect.
- Manufacturing errors.
- Overheating of the engine or other components.
- Random part failures.
- Recall-related issues.
- Service errors.
- Traction battery failures (xEVs).
- Worn-out parts.
- Other accidental causes (such as a water bottle acting as a lens focusing sunlight).
- Other human errors (for instance, smoking, mishandling flammable materials, incorrectly storing flammable materials or devices in the vehicle, and aggressive driving).

As a short example, **Figure 12.1** exhibits the burn patterns on a compact car determined to have been subjected to arson. The bubbled paint and bumper were most likely caused by an accelerant poured there. The lack of damage inside the engine compartment and on the undercarriage indicated that the fender and bumper damage was not caused by fire in either of those locations.

Figure 12.1 Arson example.

Reprinted from SAE Technical Paper 2006-01-0548.
© SAE International.

12.4.
Scientific Method

The scientific method is the most extensively applied in car fire analysis [12.3-12.13]. This systematic and logical methodology, commonly used to investigate natural phenomena and human-made systems, relies on empirical evidence and experimentation to answer questions. Its application in the analysis of car fires proves to be a potent tool, fostering precision, evidence-driven reasoning, and collaboration among experts, as described in the following sections.

12.4.1.
Precision
Car fires can have serious consequences, including injury, loss of life, and property damage. The scientific method ensures a rigorous and systematic investigation, minimizing errors and providing accurate results.

12.4.2.
Evidence-Based
Car fires involve complex interactions of mechanical, electrical, and chemical factors. The scientific method relies on evidence, rather than assumptions, helping investigators pinpoint the real causes.

12.4.3.
Replicability
By following a standardized process, investigators ensure that other researchers can replicate their findings. This is crucial for validating conclusions and improving safety measures.

12.5.
Step by Step

In harmony with NFPA guidelines, the analysis of a car fire cause involves a structured approach with the following steps described in the next sections.

12.5.1.
Recognition of the Need
The analysis is initiated by acknowledging the need to conduct a thorough investigation, typically aimed at identifying the root cause or causal nexus. In practical terms, this step is occasionally inherent to the investigator's role and may not be explicitly stated. However, it remains the primary step in the investigator's mental process.

12.5.2.
Problem Definition
The problem is clearly stated, outlining the specific aspects and questions the analysis must respond to. Unique considerations include fire origin, fire cause, arson investigation, and how the fire propagated.

12.5.3.
Data Collection

Relevant data are collected, considering elements such as the events leading up to the fire, the location, observable patterns, noticeable damages, accounts from witnesses, the availability of videos and photographs, and previous analytical reports, among other considerations. The collected data are called empirical data, derived from observation or experience, and can be verified or confirmed true.

12.5.4.
Data Analysis

In adherence to the scientific method, it is imperative to analyze all the gathered data. This crucial step must occur before the formulation of the hypotheses. It is essential to note that identifying, picking, and cataloging data do not constitute data analysis. The analysis relies on the knowledge, training, experience, and expertise of the individual conducting it. Assistance must be sought if the investigator lacks the proficiency to interpret a piece of data accurately. A comprehensive understanding of the significance of the data empowers the investigator to develop hypotheses grounded in evidence, rather than speculation.

12.5.5.
Hypothesis Formulation (Inductive Reasoning)

Grounded on the analyzed data, the fire investigator generates hypotheses to elucidate the previously defined problem, such as determining the origin or responsibility for the fire incident. This process, called inductive reasoning, involves crafting hypotheses exclusively from the empirical data gathered through observation. These hypotheses are then shaped into explanations for the event, drawing on the investigator's knowledge, training, experience, and expertise.

12.5.6.
Hypothesis Testing (Deductive Reasoning)

The fire investigator rigorously tests each hypothesis through deductive reasoning to ensure a valid and reliable conclusion. This involves comparing the hypothesis with facts and the scientific knowledge relevant to the incident. The testing process is designed to disprove or falsify the hypothesis, aiming to identify data or reasons that contradict it. This approach guards against confirmation bias, preventing reliance solely on supportive data. Hypotheses can be tested through the analytical application of scientific principles, reference to existing research, or physical experiments, with careful acknowledgment of conditions and variables. If a hypothesis is refuted, it is discarded and alternative hypotheses are tested. This iterative process continues until a hypothesis aligns uniquely with the facts and scientific principles. If no hypothesis withstands deductive scrutiny, the issue remains undetermined. However, before reaching this situation as the conclusion, adjustments or iterations in applying the methodology are usually made, ensuring a more nuanced and adequate outcome.

12.5.7.
Final Hypothesis Selection

The results of hypothesis testing are evaluated, and, ideally, a final hypothesis is selected that aligns most closely with the observed data and experimental outcomes. In some cases, even though the exact cause may not be defined, at least the most likely cause (or causes) can be characterized. After rigorous testing, the investigator reviews the entire process to ensure careful consideration of all credible data and the elimination of feasible alternate hypotheses. Failing to contemplate alternate hypotheses is a significant oversight in the scientific method. A pivotal question is, "Are any other hypotheses consistent with the data?" The investigator should meticulously document the facts supporting the final hypothesis, excluding all other reasonable alternatives.

While these steps are typically executed sequentially, as implied in **Figure 12.2**, there might be cases where iterations become necessary. Occasionally, it becomes apparent during data analysis that more details should be collected, or new insights arise during hypothesis testing, leading to a reevaluation of potential causes and the need for additional analysis tests.

Figure 12.2 Scientific method in fire analysis.

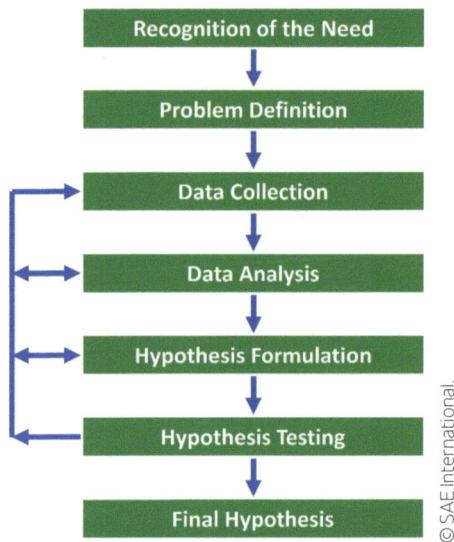

12.6.
Words of Caution

Investigators must avoid attitudes and behaviors that could compromise the analysis by preventing presumptions and being vigilant against expectation and confirmation bias.

12.6.1.
Avoid Presumption

Until data are collected, specific hypotheses should not be formed or tested. Investigators must approach fire and explosion incidents without presumptions regarding the origin, ignition sequence, cause, fire spread, or responsibility until the scientific method generates testable hypotheses that withstand rigorous testing.

12.6.2.
Beware of Expectation Bias

This phenomenon occurs when premature conclusions guide investigative processes, analyses, and conclusions without thoroughly examining all relevant data. Investigators are urged to prevent expectation bias by adhering to the proper use of the scientific method, ensuring logical and unbiased data collection and examination.

12.6.3.
Beware of Confirmation Bias

In the scientific method, hypotheses testing should aim to disprove, rather than solely confirm, them. Confirmation bias arises when investigators exclusively rely on supporting data, neglecting contradictory information. Failing to consider alternate hypotheses or prematurely dismissing seemingly contradictory data without proper analysis can lead to incorrect conclusions. Validity is established only through rigorous testing that fails to disprove the hypothesis. It involves comprehensively comparing all evidence against the proposed hypothesis to identify why it may not be valid.

12.7.
Recommended Stages

In many investigations, the initial point of the fire is not clear at the onset of the analysis, and the apparent center of the damage may not coincide with where the fire started. As seen, different velocities of burn for the available combustibles, wind influence, and other factors can deviate the flames and the damage so that the later analysis cannot solely look for the most damaged nor the central area of damage to infer that it was, in fact, the point of origin. Also, if the start spot of the fire is not correctly determined, the root cause determination will be severely impaired. Finally, presumption, confirmation bias, and expectation bias can mislead the analysis process, as discussed in applying the scientific method.

To reduce the influence of these risks, it is recommended to break down the analysis into stages, as depicted in **Figure 12.3**. First, identify where (in the vehicle) the fire started. Subsequently, proceed to ascertain the root cause or causal nexus. If relevant to the analysis scope, differentiating between accidental and intentional events should be the final stage. Ideally, apply the scientific method steps (outlined in Section 12.5) for each stage.

Figure 12.3 Recommended stages.

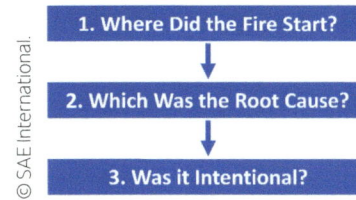

© SAE International.

1. Where Did the Fire Start?
2. Which Was the Root Cause?
3. Was it Intentional?

Consider, for instance, the scorched vehicle illustrated in **Figure 12.4**. At first glance, if no other reliable evidence exists, such as video footage documenting the fire progression from its inception, the stage "Where did the fire start?" may generate the following hypotheses:

Figure 12.4 Scorched vehicle.

Sudarshan Jha/Shutterstock.com.

H1. Engine Compartment

H2. Cabin

H3. Trunk

H4. External to vehicle

Only after selecting one of these hypotheses, the analysis should proceed to the root cause determination stage.

12.8.
Exceptions

While the scientific method is a robust and widely applicable approach to investigating fire incidents, there may be exceptions or limitations in certain circumstances. The following sections discuss some exceptions to using the standard scientific method for the analysis of a fire incident.

12.8.1.
Emergency Situations

In urgent scenarios, such as active fires or immediate life-threatening situations, investigators may need to prioritize safety and take swift actions to prevent further harm, which may, in turn, compromise the incident scene. As illustrated in **Figure 12.5**, emergency response and mitigation take obvious precedence in such cases.

Figure 12.5 Firemen in action.

Doug McLean/Shutterstock.com.

12.8.2.
Environmental Factors

Harsh environmental conditions, such as extreme weather events or hazardous materials, may hinder the complete implementation of the scientific method. Safety concerns may limit access to certain areas or compromise accurate data collection.

12.8.3.
Limited Resources

At times, investigators encounter resource limitations, be it time, personnel, or specialized equipment. When faced with such constraints, they may need to rely on available data to make informed decisions, bypassing the full scientific process. It is important to acknowledge that this approach could potentially result in less comprehensive conclusions due to the restricted depth and scope of the investigation.

12.8.4.
Legal and Practical Considerations

Legal proceedings, insurance claims, or other practical factors may require investigators to expedite their analysis or limit access to evidence. While adhering to the scientific method is the ideal approach, real-world constraints can lead to deviations.

12.8.5.
Complex Incidents

Specific fire incidents involve multiple variables, intricate circumstances, or unknown factors. In such cases, investigators may need to rely on their expertise, experience, judgment, and support of additional experts alongside the scientific method.

12.8.6.
Human Behavior and Intent

Understanding human behavior, motives, and intentions can be challenging within the strict confines of the scientific method. Investigators may need to consider or curb psychological aspects, witness statements (sometimes biased due to conflicts of interest), and other nonscientific evidence, which can impact the investigation's integrity.

12.8.7.
Historical Data

When analyzing historical fire incidents (e.g., ancient fires), obtaining direct evidence or applying the complete scientific method may be impossible. Investigators rely on indirect

clues, documented evidence, and historical context in such cases.

12.8.8.
Incomplete Information

If critical information is missing or inaccessible, conducting a comprehensive scientific analysis may be compromised. Lack of access to crucial data, witnesses, or relevant evidence can impede the application of the scientific method. This situation should be clearly stated in the documented analysis.

It is essential to recognize that while these exceptions exist, efforts should still be made to apply scientific principles and methods to the extent possible, considering each situation's unique challenges.

12.9.
Chain of Evidence

Also known as "chain of custody," chain of evidence is a crucial concept in forensic investigations in several countries (including the US and the UK), including car fire analysis. It refers to the chronological and documented history of the physical evidence involved in a case, detailing its collection, handling, storage, and analysis. Maintaining an unbroken chain of custody is essential to ensure the integrity and admissibility of the evidence in legal proceedings. In the context of car fire analysis, the chain of evidence might include the steps discussed in the following sections.

12.9.1.
Evidence Collection

The process begins with the collection of evidence from the car fire scene. This could involve items such as debris, samples of fluids, electrical components, or any other relevant materials.

12.9.2.
Documentation

Each piece of evidence must be thoroughly documented. This documentation should include the date, time, location, and the person collecting the evidence.

12.9.3.
Packaging and Labeling

Proper packaging and labeling are crucial to prevent contamination, loss, or damage to the evidence. Each item should be securely packaged and labeled with a unique identifier.

12.9.4.
Transportation

The evidence should be transported securely to the laboratory or analysis facility. It should be kept in a controlled environment during transport to preserve its integrity.

12.9.5.
Receiving at the Laboratory

Upon arrival at the laboratory, the evidence is received by forensic analysts. This step involves verifying the proof against the documentation to ensure nothing is missing or compromised.

12.9.6.
Laboratory Analysis

Forensic experts conduct their analyses on the evidence. A car fire investigation could involve examining materials for accelerants, identifying fingerprints, or performing other relevant tests.

12.9.7.
Storage

After analysis, the evidence is stored securely. Proper storage conditions are maintained to prevent deterioration or contamination.

12.9.8.
Documentation of Changes

Any changes in the status of the evidence, such as additional testing or transfers between analysts, must be documented to maintain transparency.

12.9.9.
Court Presentation

If the case goes to court, the chain of evidence is presented to establish the reliability of the findings. This includes providing a detailed account of the evidence's journey from the crime scene (or incident site) to the courtroom.

By meticulously maintaining the chain of evidence, investigators and forensic analysts establish credibility and ensure that their findings can be relied on in legal proceedings. This process is vital to uphold the criminal justice system's accuracy, reliability, and fairness standards.

References

[12.1]. National Fire Protection Association, "Standard for Professional Qualifications for Fire Investigator—NFPA 1033," 2022.

[12.2]. Okes, D., *Root Cause Analysis*, The Core of Problem Solving and Corrective Action, 2nd ed. (Milwaukee: ASQ Quality Press, 2019), ISBN:0873899822.

[12.3]. Armstrong, J. and Green, K., *The Scientific Method: A Guide to Finding Useful Knowledge* (Cambridge, UK: Cambridge University Press, 2022), doi:https://doi.org/10.1017/9781009092265.

[12.4]. Barnett, G., *Vehicle Battery Fires: Why They Happen and How They Happen*, SAE International R-443 Book (Warrendale: SAE International), 2017, ISBN:978-0-7680-8143-5.

[12.5]. Chang, M., *Principles of Scientific Methods* (Boca Raton: Chapman & Hall, 2014), ISBN:9781482238099.

[12.6]. De Santis, T., Adams, C., Molnar, L., Washington, J. et al., "Motor Vehicle Fire Investigation," SAE Technical Paper 2008-01-0555 (2008), doi:https://doi.org/10.4271/2008-01-0555.

[12.7]. Icove, D., Haynes, G., and De Han, J., *Forensic Fire Scene Reconstruction*, 3rd ed. (Boston: Pearson, 2012), ISBN-13: 9780132605779.

[12.8]. Icove, D. and Haynes, G., *Kirk's Fire Investigation*, 8th ed. (New York: Pearson, 2018), ISBN 10:0-13-423792-7 and ISBN 13: 978-0-13-423792-3.

[12.9]. National Fire Protection Association, "Guide for Fire and Explosion Investigations—NFPA 921," 2021 edition, 2021.

[12.10]. Noon, R., *Forensic Engineering Investigation*, 1st ed. (New York: CRC Press, 2000), doi:https://doi.org/10.1201/9781420041415.

[12.11]. Scheibe, R., Shields, L., and Angelos, T., "Field Investigation of Motor Vehicle Collision-Fires," SAE Technical Paper 1999-01-0088 (1999), doi:https://doi.org/10.4271/1999-01-0088.

[12.12]. Shields, L. and Scheibe, R., "Computer-Based Training in Vehicle Fire Investigation Part 1: Ignition Sources," SAE Technical Paper 2006-01-0547 (2006), doi:https://doi.org/10.4271/2006-01-0547.

[12.13]. Shields, L. and Scheibe, R., "Computer-Based Training in Vehicle Fire Investigation Part 2: Fuel Sources and Burn Patterns," SAE Technical Paper 2006-01-0548 (2006), doi:https://doi.org/10.4271/2006-01-0548.

Chapter 13

Data Collection and Analysis

13.1.
Introduction

This chapter focuses on the pivotal aspect of data collection, from safety considerations and preliminary data assessment to on-site preparation; the chapter lays out a road map for investigators for whom it is targeted.

The chapter emphasizes safety by covering essential aspects such as PPE, hazard identification, and compliance with regulations. It delves into preliminary data and preparation, setting the stage for a structured analysis. The chapter further explores the essential elements of a field kit and emphasizes the importance of checklists and witness interviews.

This chapter unfolds the nuances of on-site field analysis and vehicle analysis. Covering safety reminders, documentation protocols, and analysis of specific systems and components, these sections provide a comprehensive guide.

The chapter concludes by delving into postanalysis reflections and considerations, urging investigators to assess gathered evidence and explore avenues for deeper understanding. This condensed overview encapsulates the meticulous approach necessary for practical vehicle fire analysis during the data collection phase [13.1-13.3].

13.2.
Safety

Safety concerns must be considered during field and vehicle analysis to ensure the investigation personnel's well-being and integrity [13.4-13.6]. Some essential safety concerns are discussed in the following sections.

13.2.1.
PPE

Prioritize your well-being with comprehensive PPE coverage, adapting to the risks inherent in the environment and vehicle you are investigating. Adequate clothing, boots, gloves, safety helmets, protection glasses, and masks are some common examples. Tailor your respiratory protection based on the situation. For instance, a particulate matter mask may suffice sometimes, but more sophisticated devices are indispensable for scenarios involving lingering toxic fumes.

13.2.2.
Fire Scene Hazards

The fire scene itself can pose various hazards. In addition to deserted or wild places, unstable structures, power lines, and hazardous materials within the vehicle, investigators should also exercise caution regarding potential animal threats. Rodents, snakes, and other wildlife might be present or concealed in or around the wreckage. Investigators should also be cautious of sharp objects, broken glass, fluid spills, and other physical hazards.

Evaluate the place before approaching the vehicle. Look for structural instability, electrical hazards, fuels, or other dangers. Avoid relying solely on the park brake system as it may fail during examination. Instead, secure the wheels by using physical obstacles (chocks). Remain aware that a burned-out car might harbor weakened structural components, susceptible to collapse during the investigation.

13.2.3.
Fire Residue Exposure

The residues left after a fire may contain harmful substances, including toxic chemicals, corrosive materials, heavy metals, and other contaminants. Investigators must avoid direct contact with these residues and take precautions to prevent skin exposure, inhalation, or ingestion.

13.2.4.
Explosive Hazards

Vehicles have components that can explode or release pressure during a fire investigation. Investigators should know fuel and gas tanks, airbag inflators, seat belt pretensioners, gas struts, and other potentially explosive elements. Proper precautions should be taken to mitigate these risks.

13.2.5.
Electrical Hazards

Even after a fire is extinguished, a vehicle's electrical system, whether 12 V or the high voltage in electric and hybrid cars, can pose a threat. Turn off and disconnect all power sources before the investigation, exercising caution around exposed wires and potentially energized electrical components.

You should only examine damaged xEVs if you are qualified to scrutinize high-voltage systems and are familiar with the particular vehicle.

13.2.6.
Environmental Considerations

Car fires can release pollutants into the environment. Investigators should be mindful of the potential impact on air, soil, and water quality. Proper containment and disposal of hazardous materials must follow environmental regulations.

13.2.7.
Communication and Emergency Planning

Car fire investigators should work in pairs or teams to enhance safety. Establishing clear communication protocols and having an emergency response plan are crucial in case unexpected hazards arise during the investigation.

13.2.8.
Training and Experience

Investigators should have proper training and experience in fire scene analysis. Understanding fire behavior, forensic techniques, and safety protocols is essential to conducting a thorough and safe investigation.

13.2.9.
Legal and Regulatory Compliance

Investigators must be familiar with local, state, and federal regulations governing fire investigations, and safety standards, such as OSHA in the US, are crucial [13.7]. Adhering to legal and regulatory requirements ensures the investigation is conducted responsibly, safely, and ethically.

13.2.10.
Decontamination

Decontaminate yourself and your equipment after leaving the scene. Remove contaminated clothing and wash thoroughly. Prevent cross-contamination by adequately disposing of gloves and cleaning tools.

13.3.
Preliminary Data

Collect all available information before preparing a field analysis, typically in document and image formats. This includes but is not limited to police reports, fire department reports, analysis reports, vehicle details, witness statements, photographs, videos, recalls, service orders, and maintenance records.

Conduct a preliminary analysis of this collected material, refraining from prematurely pinpointing the root cause. Maintain an open mindset during this phase, focusing on understanding the essential information required for the subsequent preparation and field analysis. Address the following key considerations:

- Urgency of Field Analysis: Assess whether the field analysis is urgent, necessitating immediate attention. Timeliness is crucial as delaying the analysis may result in the loss or disturbance of significant evidence.

- Vehicle Location: Determine if the vehicle (or vehicles) is still at the scene or has been relocated. Preferably, analyze the car at the actual location of the incident.

- Risk of Evidence Loss or Deterioration: Evaluate whether there is an imminent risk of evidence loss or deterioration at the scene or vehicle level.

- Documentation of Firefighters' Actions: Ascertain if the actions of firefighters during extinguishing and victim rescue have been thoroughly documented.

- Resolution of Conflicting Information: Identify any conflicting information and determine if it can be clarified before proceeding with the field analysis.

- Safety Risks or Hazards: Research known safety risks or hazards associated with the scene or vehicles.

- Need for Special Resources: Determine whether unique resources are necessary for the field analysis.

- Regulatory Considerations: Explore unique local or state investigation regulations.

- Vehicle Modifications and Accessories: Investigate whether the vehicle has undergone modifications or if any accessories have been installed.

- History of Collisions and Failures: Determine if the vehicle has been involved in previous collisions or experienced failures.

- Maintenance and Service Records: Check for the availability of maintenance and service records for the vehicle.

- Recalls and Similar Events: Investigate if the vehicle is subject to any recalls and explore similar events reported on the Internet.

- Internal Incident Reports: Check whether your institution or company has records of similar incidents for additional insights.

13.4.
Preparation

After a brief analysis of the available information and initiating requests for additional data or clarification of contradictions, take a moment to strategically organize activities before heading to the fire scene and the vehicle(s)' storage location. Document your action plan, engage with relevant stakeholders, and gather all essential equipment, tools, and gear.

In preparation for accident reconstruction scenarios, consider leveraging historical images of the scene, available on platforms like Google Maps™ and Street View™. This approach may unveil changes in road signs or other conditions before and after the incident, providing valuable context for your analysis. By incorporating these preparatory steps, you enhance the efficiency and effectiveness of the fieldwork.

13.5.
Field Kit

Tailor your "field kit" to the specific incident type and details gleaned from preliminary data analysis. Consider the following list as a suggestion:

- Checklist and a clipboard for easy note-taking.

- Relevant documents and files.

- Safety-related drawings: Include safety-related drawings of the vehicle (e.g., location of airbag inflators and MSD), accessible through a phone app in some countries.

- Pen or pencil for note-taking.

- Wheel chocks to ensure vehicle immobilization (see **Figure 13.1**).

- Digital camera, spare memory card, and extra batteries.

- Polarized lens: Useful to enhance contrast for tire braking marks.

- Small light source for areas inaccessible to camera flash or phone light.

- Boroscope to explore inner engine areas or secluded spots. Inexpensive models exist that can be connected to your phone.

- Small mirror for examining hard-to-reach areas.

- Digital audio recorder or phone app for voice, instead of paper notes.

- Regular brush for cleaning and sorting debris. Consider employing a steel brush to enhance the legibility of chassis, engine, or transmission numbers.

- Small plastic bags for collecting and identifying samples.

- Measuring tools, for example, a measuring tape (metallic or laser-based) and a caliper for precise measurements.

- Miscellaneous tools including screwdrivers, pliers, and a fixed key set.

- A thermography camera or imaging thermometer, particularly for recent fires or xEV incidents (see **Figure 13.2**).

- Voltmeter, especially for xEV analysis.

- Suitable jack for lifting purposes.

- Adequate clothing and PPE.

- Rags.

- Hand sanitizer if facilities are unavailable near the fire scene.

Figure 13.1 Wheel chock.

hamza mahsun atmaca/Shutterstock.com.

Figure 13.2 Thermal imaging camera.

Dario Sabljak/Shutterstock.com.

13.6. Checklist

If your institution does not have a standard form or checklist, prepare your own. It is a fundamental tool, as represented in **Figure 13.3**, and worth the time it takes to prepare and complete.

Consider incorporating the following topics into your car fire analysis checklist:

- Vehicle Identification: Capture details such as make, model, license plate, chassis number (VIN), fabrication and model year, color, and fuel type.

- Vehicle Details: Document the last known mileage, engine number, and transmission number.

- Contact and Analysis Participants' Information: Include the name and contact information of the owner, driver, vehicle occupants, and all personnel present at the analysis.

Figure 13.3 Fire analysis checklist.

Kmpzzz/Shutterstock.com.

- Pre-Identification of Analysis Focus: Identify which parts, systems, or scene features require more detailed analysis.

- Environmental Conditions: Record weather conditions during the incident and note road conditions.

- Road Signage: Assess the presence and state of vertical and horizontal road signs.

- Surveillance Cameras: Inquire about their existence (requesting images for later analysis).

- Scene Drawings and Pavement Marks: Draw the scene and document the eventual presence of tire braking or metal gouging marks on the pavement or soil.

- Abnormal Materials: Identify and register the location of abnormal materials around the vehicle or inside its cabin or trunk. Search for potential evidence of accelerator containers or materials.

- Collision Sequence: Draw the reported collision sequence if applicable.

- Vehicle Drawing: Create a drawing of the vehicle, considering a segmented aerial view for easy cross-referencing during subsequent analysis.

- Tires and Wheels: Describe the type, condition, potential mismatch of tires and wheels, and eventual lack of some wheel lugs.

- Fluid Levels: Record the levels of oil, transmission fluid, radiator fluid, and steering system fluid, if possible.

- Fuel System Status: Evaluate the fuel system from the filler cap to the tank and engine.

- Exhaust System: Inspect the exhaust system, including turbocharger, EGR, DPF, catalytic converter, and SCR, if equipped.

- Airbags and Seatbelt Pretensioners: Assess their status.

- Ignition and Security Features: Verify the presence of ignition keys in the cylinder, assess whether doors and the trunk were locked, and look for signs of tampering.

- Maintenance: Check for information on oil/filter service stickers.

- Odometer Reading: Note the odometer reading, if feasible.

- Collision Marks and Evidence: Document the height and location of collision marks, if applicable.

- Windows: Assess the condition and position of glass and windows and their type (i.e., electric or manual).

- Missing Parts: Check for any missing parts, as well as signs of whether they were removed before or after the fire.

- Consider the following list of questions to be addressed to the witnesses:

  - Who was driving the vehicle at the time of the incident?

  - How long was the trip until the fire was noticed? Alternatively, how long was the vehicle idling before the fire was detected?

  - Did the driver or witnesses notice any electrical or mechanical failures before the fire?

  - Were there any warning lights or unusual sounds preceding the incident?

  - Did the driver or witnesses observe any abnormal behaviors, sounds, or smells from the vehicle before the incident?

  - How was the fire initially detected?

  - How did the fire evolve?

  - What measures were taken to extinguish the fire, and how was it ultimately extinguished?

  - Did the vehicle travel close to any wildfire?

  - When was the vehicle last serviced, and were there any recent maintenance activities?

- Has the vehicle experienced any collisions in the past?

- Can the witnesses specify which accessories were installed on the vehicle and when they were added?

- Were any components removed or altered after the incident?

- Does the witness have additional photo or video recordings of the fire?

13.7.
Field Analysis

13.7.1.
Safety Reminder

Before commencing any field activity, consider the safety concerns listed in this chapter. Be mindful of the potential rekindling of fires, especially in the case of xEVs with stranded energy in traction batteries or supercapacitors.

13.7.2.
Fire Scene Preservation

Secure the scene to prevent unauthorized access and refrain from disturbing evidence. Preserve crucial elements such as fire patterns, ignition point, and burn damages. Recognize the uniqueness of each fire scene, adhering to established protocols, and prioritizing safety throughout the evidence-gathering process.

13.7.3.
Documentation

Commence the documentation process by sketching the scene if deemed necessary. Before any interaction, photograph the area, encompassing the vehicle's condition, location, and surroundings. Employ a systematic approach, commencing with panoramic views and progressing to sequential, close-up images. This method facilitates the development of a thorough analysis report, eliminating the need for intricate indexing while taking each picture. Allocate time for capturing detailed images and conducting analyses of the most severely affected areas or suspected initial points toward the conclusion of the documentation process.

13.7.4.
Surveillance Cameras

Take note of any closed-circuit television (CCTV) cameras or similar devices in the vicinity (refer to **Figure 13.4**). Later, request copies of relevant video footage for further investigation.

Figure 13.4 Surveillance camera on a parking lot.

Aleksandrkozak/Shutterstock.com.

13.7.5.
Evidence Collection

Identify and meticulously collect relevant samples, including glass fragments, burned materials, fluids, and containers, with the intention of potential laboratory analysis, ensuring their original locations are duly documented. Exercise heightened scrutiny toward materials that might have served as the ignition source or containers for transporting accelerant fluids. Handle the collected items with utmost care to prevent any accidental smudging of potential fingerprints.

Consider that bottles, cans, jars, or other container types might have been haphazardly scattered after fluid discharge to the vehicle. It is conceivable that plastic containers were

strategically placed near the car, anticipating their consumption by flames, leaving behind no discernible trace. It is worth noting that arsonists are unlikely to take the containers with them from the scene.

13.7.6.
Assessment of Interventions

Conduct a thorough examination of the scene, vehicle, and its spatial arrangement to identify any indications of alterations occurring before or after the fire. Scrutinize for inconsistencies that could suggest deliberate actions leading to the fire. Additionally, weigh the likelihood of such modifications being attributed to the actions of first responders or firefighters involved in rescue operations or firefighting efforts.

13.7.7.
Assessment of the Environment for Deliberate Acts

Consider whether the location of the fire provided conducive conditions for intentional damage. This involves analyzing the environment in terms of isolation or human activity, that is, inquire whether the area in question was remote, had low foot traffic, or was characterized by sparse human presence. Locations with these characteristics may become more likely targets for vandalism or intentional fires.

Also, consider the time of the incident, mainly if it occurred early morning. Fires initiated during this period are more challenging to notice due to the usual absence of potential witnesses, favoring criminal activities.

Consider delving into historical records of riots or vandalism near the location, as illustrated in Figure 13.5. This exploration might add valuable context to the assessment, shedding light on potential correlation with the case under analysis.

13.7.8.
Before You Leave the Scene

Review your checklist to ensure comprehensive evidence capture and recording. Confirm the collection and identification of all necessary samples. Verify the completion of activities by all team members.

13.8.
Vehicle Analysis

In this section, whether at a fire scene or a different location, crucial details are explored. From safety reminders to photographic protocols and various system examinations, it provides a comprehensive guide for investigators. It addresses key aspects such as vehicle identification, body condition, assessment of the main systems, and intentional acts. Serving as an indispensable road map, it ensures a meticulous exploration of every facet of vehicle analysis.

Figure 13.5 Vehicle burning during protest.

bgrocker/Shutterstock.com.

13.8.1.
Safety Reminder

Before any contact with the vehicle, observe the following safety measures:

- Approach the vehicle at 45° from its pathway, as it might suddenly move.

- Secure at least one wheel with a suitable chock.

- Apply the parking brake and place the gear lever in park, if feasible.

- If feasible, turn the ignition key off and store it securely. Store proximity keys at least 15 ft (5 m) away from the vehicle.

- Disconnect the 12 V (or 24 V) battery, ensuring no reconnection risk. Ideally, cut the cable in two points.

- Remove and securely store the MSD plug for xEVs equipped with it. Ensure unanimous consent among team members before retrieval. Disconnecting the MSD does not discharge the traction battery; it only cuts power to most high-voltage electronics. Unplugging the MSD will not deactivate the low-voltage system (i.e., 12 V or 24 V). Therefore, airbags and other pyrotechnic devices will remain energized until the low-voltage battery is disconnected.

- For xEVs without an MSD, if you are qualified, use the standard method to disconnect the high-voltage system by cutting a specific cable in two points. Consult vehicle-specific safety information and employ a voltmeter to verify zero energy before touching related wires and connectors. In cases of uncertainty, assume all high-voltage points are energized.

- When handling xEVs, follow a safety practice of using only one hand to touch potentially energized parts, such as battery cells. This precaution helps prevent inadvertently closing an electric loop through the other hand or other body parts that might be in contact with the chassis or other circuits.

- For xEVs, unusual odors and eye, nose, throat, or skin irritation might indicate the leakage of harmful gases from the traction battery. Stop the analysis, ensure the battery is not undergoing thermal runaway, and only resume with adequate PPE.

- For xEVs, sparks, smoke, bubbling, or hissing sounds might indicate a thermal runaway. Cease the analysis, confirm battery stability, and only continue with suitable PPE.

- Remain cautious regarding structural parts that may collapse along the investigation. Hood struts can no longer be trusted, nor can the door check straps.

- If the car is elevated at any moment, be sure to use adequate equipment and PPE and respect safety regulations, such as OSHA in the US.

13.8.2.
Photography Protocol

Commence photographing the vehicle systematically, covering all sides and corners in an easily followable sequence. This organized approach will aid in efficient reference when compiling your report. Include shots of the roof, hood(s), and lower regions (as much as possible) before any elevation occurs. Begin with a focus on less damaged areas. Intersperse zoomed-out photos to progressively clarify their position in the vehicle when capturing finer details. For example, when photographing a tire feature, include a photo that distinctly identifies the detailed tire (e.g., front right). This methodology eliminates the need for individual photo indexing during analysis, resulting in quicker photo retrieval and identification when composing the analysis report, ultimately saving time.

13.8.3.
Vehicle Identification

Ensure that you are indeed facing the anticipated vehicle and its key components. Despite a matching license plate, scrutinize the VIN inscription as it might reveal discrepancies or signs of tampering. While uncommon, instances of fraud may involve the upfront replacement of the engine, emphasizing the need for thorough verification.

13.8.4.
Vehicle Body

Conduct a detailed examination of the vehicle body to identify critical indicators. Check for intentional door openings and assess burn patterns on the cabin envelope's inner and outer sides. Document the condition and positioning of glass and windows; pay attention to the attachment of windshield remnants to the upper or lower frame.

Additionally, scrutinize the vehicle for signs of collisions or extensive damage unrelated to the fire, evaluating their alignment with existing reports. Lastly, analyze fire patterns on the vehicle's surfaces to gain valuable insights into the direction and intensity of the fire [13.8-13.13].

13.8.5.
Fuel System

Conduct a detailed inspection of the vehicle's fuel system, including an assessment of the fuel type (gasoline, diesel, CNG, etc.) and its influence on the fire. Additionally, scrutinize the fuel tank for any visible damage that may have contributed to the incident. Identify potential issues such as leaks or compromised components, considering the challenge of determining prefire damage, particularly with the prevalent use of plastic parts.

13.8.6.
Engine and Transmission

Inspect whether the carter, engine, or transmission walls exhibit fractures and assess their secure attachment. Verify the structural integrity of these components to ensure they remain adequately connected.

13.8.7.
Exhaust System, Catalytic Converters, Turbochargers, and EGR

Examine for any indications of notable disconnections, fractured envelopes, deformations that might have caused overheating in nearby components, remains of foreign material (e.g., grass or rags), and so forth [13.14].

13.8.8.
Electric system

Retrieve defect trouble codes and stored information from the event data recorder, airbag control module, tachograph, and tracking device, as applicable, relevant, and feasible. In certain instances, surviving modules may need removal and energizing in a laboratory or by a part supplier. Sometimes, information can be retrieved through authorized dealer equipment [13.15].

Check for blown fuses, acknowledging that this assessment may be pointless in cases of significant electrical damage. Identify any darkened electric connectors that could be linked to the fire origin, recognizing that flames often cause damage to connectors and harnesses. Assess the condition of the high-power harness, including the starter motor and alternator feeds.

Verify the proper fastening of the (12 V or 24 V) battery and the integrity of all battery connections, considering the possibility of disconnection during firefighting efforts.

Examine electronic modules for indications of water infiltration or corrosion unrelated to firefighting efforts or subsequent exposure to the environment. Exploring evidence of water intrusion, such as blackened copper filaments on wires connected to the module suspected to be the fire origin, may be a feasible investigative approach in partially burned vehicles.

13.8.9.
Evidence of Short Circuits

On close examination, copper beads are usually discerned on portions of the vehicle wiring, particularly when flames have significantly affected the vehicle. The conclusive determination of whether these traces are linked to the root cause or represent secondary damage often unfolds in later stages of analysis. In the interim, focus on diligently identifying wires exhibiting these signs. Pay attention to the presence of copper beads, bubbled wire insulators, hardened copper wires, and darkened connectors. Notably, the complete absence of copper beads indicates either a prior disconnection of the battery or a previous depletion of its energy.

13.8.10.
Braking System

Look for indicators of overheating, such as a bluish hue on brake drums or discs, as well as glazed or cracked brake shoes. Glazing relates to a highly polished appearance on the lining material surface, which might result from overheating. Yet, its surface may also have a black, flaky material accumulation. Take also into account that depending on the brake liner material, it can have surpassed the autoignition temperature of the corresponding tire, without glazing its surface. Cracked shoe liners, lining material falling off the shoe, and charred liner appearance are also relevant. The recommended maintenance and inspection practice RP 607C, *Preventive Maintenance and Inspection of S-Cam Foundation Brakes*, created by the TMC, provides valuable details on S-cam foundation brake failures and related evidence [13.16].

Examining the braking system is more significant for large trucks and buses, where drivers may not readily detect braking issues and could continue driving over extended periods. **Figure 13.6** shows an instance of overheated brake shoes from a study by Parrott and Stahl [13.17]. A comparative analysis between the left- and right-side brake shoes of the same axis can prove fundamental to identifying if the overheating of one of the shoes triggered the fire. Due to the large mass of the wheel drum in trucks and trailers, if the brake shoe triggered the fire, the grease inside this wheel (on its bearings) might have been consumed. In contrast, grease will usually remain visible for the wheels of the opposed side (that did not experience a shoe overheating), even if its tire is consumed during the fire, as secondary damage on this side.

Check for potential damage to brake system lines and consider whether the fire originated from a tire. Vigilance in assessing these elements contributes crucial insights into the braking system's condition, especially when prolonged operation may have masked underlying issues.

Figure 13.6 Burnt brake shoes.

13.8.11.
Potentially Intentional Acts/Arson

Be vigilant for unconventional signs that may suggest intentional acts or arson. Look out for personal or unusual objects in unexpected locations, remnants of matches, matchboxes, or combustible materials. Consider the possibility of the fire being a smokescreen to conceal another crime, such as murder.

Car fires can be initiated by placing ignited newspaper or rags near a seat while leaving a door or window open to ensure oxygen flow. Even in such cases, some remnants may withstand the fire.

Potential traces of accelerants and arson evidence include irregular burns on the rubber door or window seals, unusual bubbles or dripping marks on painted sheet metal parts, bubbles or burned paint on the side that remained cold, and incoherent burns on closely painted and plastic components (refer to **Figure 13.7**). Unexpected combustible materials below the engine or tank, multiple burn marks in distant vehicle spots without communication, and the apparent origin of the fire in a place without fuel/hydraulic lines or electric components are also notable indicators.

Figure 13.7 Incoherent burnt suggestive of arson.

Anton Vakhrushev/Shutterstock.com.

However, be mindful of factors that might create misleading patterns:

- The miscellaneous paint layers of a typical vehicle leave different residues with variable appearances based on burn temperature, ventilation, exposure to humidity, and other factors.

- Structural reinforcements in certain sheet metal parts (e.g., the engine hood) often create distinct patterns due to temperature variations during the fire.

- The use of water and other chemicals during firefighting.

- The opening of doors, hoods, or trunks during firefighting or victim rescue activities.

- The evolution of corrosion colors and patterns over time, especially on steel parts, even in just a few hours.

13.8.12.
Airbag and Restraint Systems

Check for deployed airbags and assess the condition of seat belts—note if they are intact or potentially cut for victims' extrication. Additionally, investigate whether seatbelt pretensioners have been activated.

13.8.13.
Missing Parts

Specific vehicle components may detach during a fire, such as the gas cap, side mirrors, door handles, headlamps, grills, and more. Expect to find some of their remnants in the debris surrounding the car if they were not removed before the fire. Remember that certain plastic parts may completely burn away, while others might be ejected or displaced due to metallic springs.

Conversely, arsonists often remove high-value items before initiating a fire or replace them with more economical parts, presuming the fire will obliterate any evidence of their actions. However, a meticulous examination can reveal indicators, such as the absence of the original multimedia system and missing tire lugs (indicating tire replacement). Examining tire tread remains facing the ground may expose worn-out or "junk" tires.

13.8.14.
Other/Miscellaneous

Consider the presence of a burnt cell phone charger—could it have been the catalyst for the fire? Also, consider the role or malfunction of any aftermarket modifications or accessories. Examine the glove box or storage bins for the availability of car documents; retrieve revision and service information if present. Look for signs or inconsistencies that may be explained by first responder extrication activities, as illustrated in **Figure 13.8**, firefighter actions, the passage of time, or third-party actions (e.g., bystanders and police). Factor in potential damages and missing parts due to suboptimal storage conditions or towing operations (if the vehicle was relocated from the original fire scene).

13.8.15.
xEVs

Chapters 7 and 8 provided more details related to these types of vehicles. It is advisable to cross-reference the latest information with "Alternative Fuel Vehicles Safety Training Program Emergency Field Guide" by the NFPA to ensure comprehensive coverage for the specific vehicle in question [13.2]. Additionally, take into account the following list of unique scenarios and the associated risks.

13.8.15.1.
High-Voltage Components

xEVs have high-voltage systems, which can shock and electrocute investigators. Proper safety protocols must be ensured when analyzing the car and handling damaged components.

13.8.15.2.
LIBs

Most of these vehicles use LIBs, which can contain stranded energy. This energy may allow

Figure 13.8 Extrication damaging doors.

ChiccoDodiFC/Shutterstock.com.

them to catch fire, explode if damaged, and rekindle the fire. Investigators should approach battery packs with caution and avoid puncturing or short-circuiting them. Also, when handling individual cells, consider the possibility of them suddenly shorting and exploding, as there have been reports of such occurrences [13.18].

13.8.15.3.
Thermal Runaway

LIBs can also experience thermal runaway, where internal reactions lead to rapid heating and fire. Due to stranded energy, this event can happen again hours, days, or even weeks after the initial fire. Also, if the thermal runaway gases are not immediately ignited, they can accumulate (in the vehicle cabin or a garage), creating explosion risks. Investigators must be aware of these phenomena and take necessary precautions.

13.8.15.4.
Chemical Hazards

Battery fires release toxic gases and chemicals. Proper protective gear is essential to prevent exposure to harmful substances during investigations.

13.8.15.5.
Supercapacitors

Some xEVs incorporate supercapacitor modules in their electric systems, that is, banks of these electronic components that act like large-value capacitors are able to handle prominent charge/discharge current peaks efficiently. Like lithium battery modules, they can store charge, and therefore high voltages, for an extended period (several days) after the vehicle has been turned off. They also have numerous toxic substances, and investigators must be aware of the associated dangers.

13.8.15.6.
Hidden Fires

Damaged battery cells, supercapacitors, or wiring may continue to smolder even after visible flames are extinguished. Thorough inspections are necessary to identify hidden fire sources. Consider utilizing a thermography camera to verify the absence of developing hot spots.

13.8.16.
Apparent Fire Dynamics

Can you preliminary discern the fire dynamics—its origin, initiation, spread, and the factors influencing its behavior, such as wind or accelerants? While immediate answers may be elusive, pondering these questions while analyzing the vehicle sparks awareness, fostering your observation and documentation of relevant details.

13.8.17.
Before You Leave the Vehicle

Before concluding your investigation, carefully review the checklist to ensure all essential evidence has been captured and documented. Collect and identify necessary samples and verify that all team members have completed their tasks.

Return any keys and car documents to the owner or authorities. When there is a notable risk of needing to reexamine the vehicle, take measures to ensure its preservation until that need is conclusively resolved. In situations involving xEVs, where residual energy in the battery or supercapacitors may exist, it is crucial to ensure the implementation of effective isolation measures and that careful handling of the vehicle remains will be exercised.

13.9.
Postfield and Vehicle Analysis

Reflect on the evidence gathered or analyzed during field or vehicle assessments. Did it unveil new inconsistencies, highlight the necessity for additional information, or indicate the need for a more profound analysis? If available documentation falls short of providing comprehensive insights, contemplate exploring a similar vehicle in optimal conditions.

Moreover, consider collaborating with or seeking guidance from experts in relevant fields, such as automotive engineers, electrical engineers, or fuel specialists. Their insights can offer valuable perspectives and contribute to a more thorough understanding of the intricacies.

References

[13.1]. National Fire Protection Association, "Guide for Fire and Explosion Investigations—NFPA 921," 2021.

[13.2]. National Fire Protection Association, "NFPA's Alternative Fuel Vehicles Safety Training Program Emergency Field Guide," 2018, ISBN:978-1455912742.

[13.3]. Icove, D. and Haynes, G., *Kirk's Fire Investigation*, 8th ed. (New York: Pearson, 2018), ISBN 10:0-13-423792-7 and ISBN 13:978-0-13-423792-3.

[13.4]. Brach, M., *SAE International's Dictionary of Vehicle Accident Reconstruction and Automotive Safety*, SAE International R-556 Book (Warrendale: SAE International, 2023), ISBN:978-1-4686-0594-5.

[13.5]. SAE International Ground Vehicle Standard, "Hybrid and EV First and Second Responder Recommended Practice," SAE Standard J2990, J2990_201907, July 2019.

[13.6]. SAE International Ground Vehicle Standard, "Hybrid and Electric Vehicle Safety Systems Information Report," SAE Standard J2990/2, J2990/2_202011, November 2020.

[13.7]. Occupational Safety and Health Administration (OSHA), "Fire Safety Standards," accessed January 2024, https://www.osha.gov/fire-safety/standards.

[13.8]. Brach, M., Brach, R., and Mason, J., *Vehicle Accident Analysis and Reconstruction Methods*, 3rd ed., SAE International R-516 Book (Warrendale: SAE International, 2022, ISBN:978-1-4686-0345-3.

[13.9]. De Santis, T., Adams, C., Molnar, L., Washington, J. et al., "Motor Vehicle Fire Investigation," SAE Technical Paper 2008-01-0555 (2008), doi:https://doi.org/10.4271/2008-01-0555.

[13.10]. Icove, D., Haynes, G., and De Han, J., *Forensic Fire Scene Reconstruction*, 3rd ed. (Boston: Pearson, 2012), ISBN-13: 9780132605779.

[13.11]. Noon, R., *Forensic Engineering Investigation*, 1st ed. (Boca Raton: CRC Press, 2000), doi:https://doi.org/10.1201/9781420041415.

[13.12]. Scheibe, R., Shields, L., and Angelos, T., "Field Investigation of Motor Vehicle Collision-Fires," SAE Technical Paper 1999-01-0088 (1999), doi:https://doi.org/10.4271/1999-01-0088.

[13.13]. Shields, L. and Scheibe, R., "Computer-Based Training in Vehicle Fire Investigation Part 2: Fuel Sources and Burn Patterns," SAE Technical Paper 2006-01-0548 (2006), doi:https://doi.org/10.4271/2006-01-0548.

[13.14]. Morse, T., Cundy, M., and Kytomaa, H., "Vehicle Fires Resulting from Hot Surface Ignition of Grass and Leaves," SAE Technical Paper 2017-01-1354 (2017), doi:https://doi.org/10.4271/2017-01-1354.

[13.15]. Barnett, G., *Vehicle Battery Fires: Why They Happen and How They Happen*, SAE International R-443 Book (Warrendale: SAE International, 2017), ISBN:978-0-7680-8143-5.

[13.16]. TMC, "RP 607C, Preventive Maintenance and Inspection of S-Cam Foundation Brakes," RP 607C Appears in the Technology & Maintenance Council's (TMC) 2024-2025 Recommended Practices Manual, (703) 838-1763, accessed April 2024, https://tmc.trucking.org.

[13.17]. Parrott, K. and Stahl, D., "Case Studies of Parking Brake Fires in Commercial Vehicles," SAE Technical Paper 2013-01-0207 (2013), doi:https://doi.org/10.4271/2013-01-0207.

[13.18]. Denver 7 ABC News, "Close Call: Lithium-Ion Battery Explodes in Adams County Fire Investigator's Face," accessed April 2024, https://www.denver7.com/news/local-news/close-call-lithium-ion-battery-explodes-in-adams-county-fire-investigators-face.

Chapter 14

Hypotheses Development and Testing

14.1.
Introduction

In this comprehensive exploration of fire investigation methodologies, this chapter, also aimed at the investigator, delves into the intricate process of formulating and testing hypotheses to decipher the origin, root cause, and nature of vehicular fires. The chapter unfolds a structured framework, categorizing hypotheses into three critical dimensions: start location hypotheses, root cause hypotheses, and accidental/intentional hypotheses. This systematic breakdown sets the stage for a meticulous analysis that unravels the where and why of a fire and explores the critical aspect of intentionality.

As the investigation progresses, the text transitions into developing hypotheses, where the culmination of collected data, information, and scientific knowledge converges into a set of plausible explanations for the origin and progression of the fire. The subsequent sections unfold a rigorous testing hypotheses phase, using evidence to debunk or support various causes. From scrutinizing fuel and thermal factors to probing into electrical, design, manufacturing, and service-related issues, each hypothesis must undergo meticulous examination, ushering in a logical and deductive approach to the investigative process.

The narrative further unfolds into ideal and unexpected outcomes, admitting the investigator's aspiration for a streamlined progression through hypotheses testing, ideally leading to a singular, viable conclusion. However, the chapter acknowledges the realities of investigative challenges, addressing scenarios where all hypotheses are rejected, too many persist as feasible, or a select few still need to be in contention. With insights into the intricacies of reevaluation, adjustment, and retesting, the chapter guides investigators through the dynamic nature of the scientific method.

Ultimately, this chapter culminates in the imperative of finalization of conclusions, outlining the steps to articulate the specific location and cause of the fire based on the hypothesis that best aligns with the totality of evidence. The significance of peer review and legal and regulatory considerations are underscored, emphasizing fire investigations' collaborative and meticulous nature within established standards and legal frameworks [14.1, 14.2].

14.2.
Hypotheses Categories

As discussed in Chapter 12, in several cases, it is necessary to structure the analysis into up to three stages: "Where did the fire start?", "Which was the root cause?" and "Was it intentional?" The following lists might be helpful besides the hypotheses formulated based on the data analysis. Readers are advised to peruse and customize these lists, ensuring a comprehensive consideration of all pertinent hypotheses at each stage of the analysis.

14.2.1.
Start Location Hypotheses
- Cabin.
- Engine compartment.
- Trunk.
- Cargo compartment.
- Undercarriage.
- Other/external to the vehicle.

In specific scenarios, this stage may be deemed unnecessary, especially when the initiation point of the fire is readily apparent or known. This could be the case when video footage meticulously captures the inception of the fire or when the affected area is confined due to the fire being swiftly extinguished before it could escalate further.

14.2.2.
Root Cause Hypotheses

14.2.2.1.
Typical Hypotheses
- Braking system issues.
- Collision.
- Contact of external materials with hot parts.
- Cooling system failures.
- Defective parts.
- Design errors.
- Electrical issues.
- Exhaust system issues.
- External sources.
- Faulty accessories.
- Flammable cargo.
- Fuel system leakages.
- Leakage of other combustible fluids.
- Improper accessory installation.
- Maintenance errors.
- Maintenance neglect.
- Manufacturing errors.
- Natural events, such as lightning.
- Overheating of the engine or other components.

- Random part failures.
- Recall-related issues.
- Service errors.
- Traction battery failures (xEVs).
- Worn-out parts.

Specific hypotheses can undergo refinement when more extensive evidence is available. For instance, investigating a potential fuel injector issue might involve specifying the hypothesis to focus on, such as pinpointing a leakage at the connection of the fuel injector of the second cylinder.

14.2.2.2.
Electrical or Nonelectrical Hypotheses

In certain instances, the investigative focus narrows to whether the fire's origin lies in electrical factors (like a short circuit) or nonelectrical factors (such as fuel leakage). As discussed earlier, fires initiated by electrical causes have the potential to escalate, involving various combustible materials within any vehicle. The aftermath of such fires may complicate the analysis, making it challenging to determine whether the root cause was electrical or not. A similar complexity arises in fires initiated by chemical combustion events, where flames can damage energized harnesses, leading to short circuits and leaving distinctive traces like copper beads.

Essentially, investigators may grapple with a classic chicken-and-egg dilemma, as depicted in **Figure 14.1**. Even when the evidence available does not unequivocally favor one cause category over the other, the structured analysis and sequential tests advocated in this methodology often facilitate the identification of the most probable cause based on evidence supporting one category.

Figure 14.1 Causality dilemma.

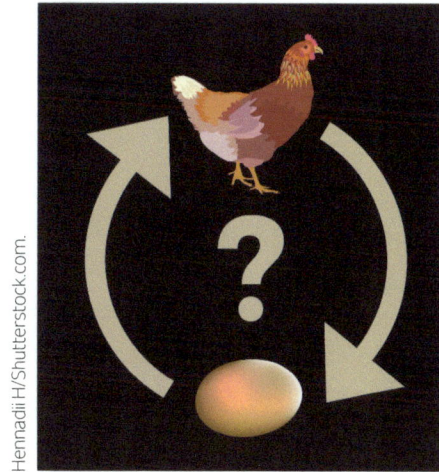

Hennadii H/Shutterstock.com.

Nevertheless, there are situations where reaching a conclusive result may prove challenging. In such instances, it is imperative for the analysis report to meticulously document the reasons for this inconclusiveness, offering a transparent account of the limitations faced during the investigation.

14.2.3.
Accidental/Intentional Hypotheses

- Accidental: The fire was not caused by intentional human action.
- Arson[1]: Deliberate setting of the vehicle on fire with malicious intent.
- Concealment of another crime: Burning the vehicle to hide evidence of another criminal activity.
- Insurance fraud: Intentionally setting the vehicle on fire to collect insurance money.

---

[1] In specific jurisdictions, such as the US, categorizing an incident as "arson" is a legal determination that a judge or jury must make. While investigators may document indicators of a deliberately set fire, they are not authorized to draw the legal conclusion of arson in their reports.

- Revenge: Deliberate vehicle targeting as revenge against the owner.
- Vandalism: Willful destruction of the vehicle through acts such as pouring fire accelerants or igniting its tires.

14.3.
Developing Hypotheses

Grounded on a comprehensive analysis of collected data, information, and scientific knowledge, employ inductive reasoning to articulate a list of hypotheses that currently appear capable of elucidating the origin and progression of the fire.

By this stage in analyzing a car fire, you likely already have a few possibilities in mind, or others may have proposed potential alternatives. In specific forensic analysis scenarios, one party might have established particular possibilities or even asserted a specific root cause as correct. Ensure all existing hypotheses are formally incorporated into your list as alternate possibilities.

Considering the safety implications of contemporary LIBs, it is advisable to include battery failure as a hypothesis in xEV fires, even if initial evidence strongly suggests otherwise. This proactive inclusion allows formal testing to rule out or substantiate this hypothesis, facilitating a comprehensive exploration of potential causes and pre-emptively addressing any concerns related to LIB failures.

14.4.
Testing Hypotheses

During this step, fire investigators employ deductive reasoning to scrutinize the set of hypotheses. They conduct logical examinations based on available data, encompassing fire patterns, fire progression, video recordings, reliable witness statements, and physical evidence. Applying the scientific method typically involves experiments, simulations, or additional analyses. However, in the context of car fire investigations, the replication of burning a similar vehicle is both expensive and generally unnecessary, given the wealth of knowledge already accessible in this field. Physical or laboratory experiments are relatively infrequent due to the costs involved and the extensive existing knowledge for evidence analysis [14.3, 14.4].

Initiate the testing process for each hypothesis against the available data and evidence, with the primary objective of rejecting it. These assessments involve posing simple questions based on the evidence and seeking reliable indications that disprove or make a hypothesis implausible within the context of the incident. For instance, if a fire ignites several hours after the vehicle has been turned off, the hypothesis of fuel leakage contacting hot engine parts can be dismissed as the engine components would not retain sufficient heat.

Once all hypotheses undergo scrutiny for potential rejection, should more than one remain viable, investigators must then conduct further tests. This time, the focus shifts to identifying evidence that strengthens the probability of a particular hypothesis being the root cause of the fire.

The subsequent lists outline typical evidence encountered in car fires, indicating hypotheses that can be ruled out (or at least considered improbable) and typical evidence that may enhance the likelihood of particular hypotheses.

14.4.1.
Evidence That Rejects Fuel/Thermal Causes

- Engine was turned off some hours before the incident (therefore, all parts were cold).
- The fire starting point had no fuel/combustible fluid sources.

14.4.2.
Evidence Suggestive of Fuel/Thermal Causes

- Fire started with the engine on or while still hot.
- Engine, transmission, turbocharger, or hydraulic steering malfunction noticed before the fire.
- Fuel smell noticed before the fire.
- The fire starting point had fuel or flammable fluid sources.
- The fire starting point has no electric modules or harnesses.
- Recent intervention on the engine, fuel system, hydraulic systems, or oil/filter change.

14.4.3.
Evidence That Rejects Electric Causes

- The fire starting point has no electric modules or harnesses.
- No evidence of short circuits nor damage to electric modules.

14.4.4.
Evidence Suggestive of Electric Causes

- Engine was turned off some hours before the incident (therefore, all parts were cold).
- Electric failures were noticed before the fire.
- The fire starting point has damaged electric modules or harnesses.

- Fire starting point had no fuel/flammable fluid sources.
- Evidence of short circuit (copper beads, hardened copper wires, or internal bubbles on wire insulators).
- Recent intervention on electric systems.
- Recent installation of electric accessories.
- PEVs during recharge.

14.4.5.
Evidence Suggestive of Design Errors

Design errors typically impact many vehicles and occasionally lead to recalls. In addition to checking for any existing recalls (in all countries where the car is sold), perform an internet search; a notable frequency of similar issues enhances the likelihood of design errors. Here are some examples of causes:

- Software lockup, keeping a load permanently energized.
- Real-life maximum temperature higher than anticipated.
- Underestimated dynamic stress of hoses or harnesses.
- Overestimated dynamic clearance, allowing rubbing wear.

14.4.6.
Evidence Suggestive of Manufacturing Errors

Manufacturing errors typically impact a smaller subset of vehicles than design errors, often becoming evident after just a few months of usage and occasionally prompting recalls. In addition to verifying any existing recalls (across all countries where the vehicle is sold), a comprehensive Internet search can be instrumental, with a notable frequency of similar issues bolstering the likelihood of manufacturing errors. Examples of potential causes are as follows:

- Insufficient torque on attached parts or the connections of fluid lines.

- Incorrect routing or attachment of hoses, leading to friction damage.

- Incorrect routing or attachment of harnesses, leading to damage and short circuits.

- Insufficient torque or paint residues on ground connections.

- Assembly of incorrect or weak parts.

14.4.7.
Evidence Suggestive of Service Errors

Typically, these errors become apparent shortly after service, with manifestations occurring within minutes to a few days. These errors often correlate with the type of service applied or the specific region of the vehicle where the intervention occurred. Here are some examples of causes:

- Rag forgotten close to the catalytic converter, after replacing the engine oil filter.

- Misconnected fuel injectors after their cleanup or replacement.

- Loosen harness or fuel lines (e.g., failing to engage a retaining clip) after a related intervention.

- Short circuits after installing aftermarket electronic modules.

14.5.
Ideal and Unexpected Outcomes

The investigator ideally anticipates a streamlined progression through the various steps, aiming

for a singular iteration. To illustrate, if the hypothesis currently under analysis (H1) successfully passes all tests, it is retained. In the event of failure, it must be promptly discarded, with the rationale for dismissal meticulously documented. Subsequently, the investigator tests the following hypothesis, H2, repeating the process until all hypotheses undergo a thorough examination. The ultimate goal is an ideal outcome where, after the analysis, only one hypothesis emerges as viable, as depicted in **Figure 14.2**, allowing for the conclusive finalization of the investigative process.

Figure 14.2 Just one hypothesis passed.

However, there are instances where all hypotheses are rejected, too many persist as viable, or more than one remain. Let us delve into a more detailed examination of what to do in such situations.

14.5.1.
All Hypotheses Rejected

When faced with the scenario depicted in **Figure 14.3**, take a moment to reassess the reliability of all pieces of evidence. Consider whether any evidence might be questionable, leading to the dismissal of the correct root cause. Additionally, explore the possibility of other hypotheses not initially included in your list being applicable. In certain situations, gathering additional data to test these new hypotheses may be necessary. If you remain at an impasse, seek input from other experts in related fields for valuable advice.

Figure 14.3 All hypotheses rejected.

inimalGraphic/Shutterstock.com.

14.5.2.
Too Many Hypotheses Remain Viable
When presented with the scenario depicted in **Figure 14.4**, it is worth considering whether your hypotheses could be refined to balance specificity and accuracy better. The available evidence may not entirely support an overly detailed approach, so finding a middle ground could enhance the overall effectiveness of your analysis.

Figure 14.4 Too many hypotheses still viable.

mstanley/Shutterstock.com.

Consider consolidating some hypotheses into broader categories. For example, if H1 focused on "fire initiated in the multimedia module" and H2 on "fire initiated in the HVAC control," you might need to merge them into a more encompassing category like "fire initiated in the panel center stack" or "fire caused by an electrical short-circuit."

Additional data may be necessary to reevaluate and test the hypotheses in certain instances. If you stay at an impasse, seeking guidance from other experts in related fields can provide valuable insights.

14.5.3.
Two or Three Hypotheses Remain Viable
At this juncture, the focus shifts to rigorously testing the remaining hypotheses with the goal of validation, instead of rejection. Does each piece of evidence affirm the plausibility of the hypothesis being tested? Furthermore, does this evidence enhance the probability of the hypothesis under scrutiny being correct?

The outcome of these new inquiries may lead to a single highly probable candidate. Alternatively, two or more hypotheses may persist, but hopefully, it becomes evident that one is more likely to be correct than the others. In such cases, the conclusion might convey something like "both hypotheses, H1 and H2, are considered possible, with H2 being more probable."

In certain instances, gathering more data may be necessary to assess further and test the hypotheses. If challenges persist, seeking input from other experts in related fields can offer valuable perspectives.

14.6.
Reevaluate, Adjust, and Retest

Be open to readjusting the course at any point. Modify or refine hypotheses based on their testing and selection results, especially in the situations described in Sections 14.5.1 and 14.5.2. Consider new information and insight that emerge at any moment while applying the scientific method. Reevaluate, adjust hypotheses as necessary, and retest. Maybe key evidence turned out to be wrong or biased. Or the evolution of the analysis may show that a chain of events might have happened, requiring the sequential occurrence of specific causes. Rewrite your hypotheses accordingly and repeat the formality of testing them against the complete set of reliable evidence.

14.7.
Finalize Conclusions

Articulate where the fire originated, aligning with the hypothesis from this category that emerged as the most plausible and harmonizes best with the evidence. Subsequently, determine the cause of the fire by selecting the hypothesis

that remains unscathed by any evidence and that all available pieces of evidence confirm this hypothesis as valid. When relevant, define whether the cause was intentional or accidental.

In instances where more than one hypothesis aligns with the available evidence, ascertain which one is more probable or identify the most likely hypotheses among those retained as potential causes.

If none of the hypotheses align with the available evidence, end your report as inconclusive.

In all scenarios, meticulously document the findings and compile a comprehensive report delineating the investigation process, evidence, hypotheses, their testing, and the ultimate conclusions.

Finally, if the analyzed vehicle belongs to a fleet, conducting thorough inspections on the remaining cars is imperative. With rare exceptions (e.g., lightning as the root cause), this is crucial because the exact root cause of the initial fire may trigger similar incidents in other vehicles. Therefore, you should strongly recommend an urgent inspection of the entire fleet in your conclusion.

14.8.
Peer Review

Ideally, and especially in your initial reports, present your findings for peer review or seek consultation with other experts in the field to validate the analysis and conclusions. This collaborative approach enhances the quality of your work and experience and contributes to the learning process for all involved, fostering a two-way knowledge exchange.

14.9.
Legal and Regulatory Considerations

Adhere to established legal and regulatory standards throughout the investigation process. Ensure that findings and conclusions are presented suitably for legal proceedings. In specific countries and roles, be prepared for potential courtroom presentations and diligently maintain the chain of custody over evidence for the required duration.

Remember that in certain jurisdictions, labeling an event as "arson" constitutes a legal decision falling under the jurisdiction of a judge or jury. While investigators can document signs of an intentionally set fire, they may lack the legal authority to declare it as arson in their reports.

References

[14.1]. National Fire Protection Association, "Guide for Fire and Explosion Investigations—NFPA 921," 2021.

[14.2]. Icove, D. and Haynes, G., *Kirk's Fire Investigation*, 8th ed. (Boston: Pearson, 2018), ISBN 10: 0-13-423792-7 and ISBN 13: 978-0-13-423792-3.

[14.3]. Gorbett, G. and Chapdelaine, W., "Scientific Method-Use, Application, and Gap Analysis for Origin Determination," Forensic Fire Analysis Institute, accessed April 2024, https://forensic-fireinstitute.com/wp-content/uploads/2018/07/scientific-method-use-application-and-gap-analysis-for-origin-determination-.pdf.

[14.4]. Freckelton, I. (Eds), *Forensic Analysis—Scientific and Medical Techniques and Evidence under the Microscope* (London, UK: IntechOpen, 2021), doi:10.5772/intechopen.92955.

Chapter 15

Case Studies

15.1.
Introduction

The cases presented in this chapter necessitated the suppression of specific details to ensure the confidentiality of individuals and the brands associated with the vehicles under scrutiny. Additionally, certain details have been intentionally modified from the actual incidents for instructional purposes.

While adhering to the scientific method wherever feasible, the presentation of each case has been streamlined; rather than providing exhaustive end-to-end narratives, a more practical and simplified approach has been adopted. Each case is introduced with the most pertinent information available before the vehicle analysis. Subsequently, the relevant hypotheses are enumerated, followed by the essential findings and the most critical images utilized in the study. The reasons behind eliminating or retaining hypotheses are discussed, culminating in the case conclusion.

Deliberately, four cases involve ICEVs: in two of the cases, electric issues were identified as the root cause; one linked to (likely) fuel leakage; and another featuring brake system failure. Notably, the last case, a BEV that experienced thermal runaway, did not reexamine the already-established root cause. Nevertheless, it furnishes readers with up-to-date information and insights applicable to analogous cases they may encounter in their investigative pursuits.

Readers are also encouraged to study other analysis documents [15.1-15.3] and images or video footage available on the Internet, from accredited sources [15.4-15.6].

15.2.
Case A: Engine Compartment Fire

Several months following its acquisition, a used subcompact hatchback car unexpectedly ignited a fire while parked in its owner's garage. In response to this incident, the owner filed a lawsuit against the vendor and the manufacturer, alleging a suspected preexisting failure as the cause of the fire. The subsequent analysis, conducted a few months after the event, aimed to ascertain the root cause of the fire to provide essential support for a judicial decision.

15.2.1.
Information Available Before Vehicle Analysis

- Flex-fuel (gasoline/ethanol) ICEV, passenger car.

- Approximately 43,000 km and two years of total use.

- Manufacturer warranty of three years still in effect.

- Third owner.

- Fire occurred four months after the most recent purchase.

- The vehicle was parked and idle for over 12 h before the observation of smoke and flames late at night.

- No pending recalls.

- No anomalies were detected before the incident.

- Firefighters extinguished the fire with water.

- Damages limited to the engine compartment.

- No significant service or installation of accessories since the car's last purchase.

- An analysis report from the manufacturer identified the root cause as a short circuit above the battery, between its positive pole and a metallic bracket.

15.2.2.
Preliminary Hypotheses

Contenders in the judicial process had already formulated certain hypotheses. Building on the available information, typical hypotheses were added, resulting in the compilation of the following lists:

15.2.2.1.
Starting Point

- S1: Inside the engine compartment.

- S2: External to the vehicle.

15.2.2.2.
Cause

- C1: Design error.

- C2: Manufacturing error.

- C3: Service error.

- C4: Deliberate.

15.2.2.3.
Intent

- I1: Intentional.

- I2: Accidental.

15.2.3.
Key Findings during Vehicle Analysis and Owner's Interview

- Main damages were concentrated in the upper region of the engine compartment.

- Neighbors helped push the car out of the garage, given the impossibility of starting the engine and concerns that flames would reach the house.

- Some external damages were caused during the vehicle removal from the garage and firefighting efforts.

- Original battery and tires were in use since the vehicle's manufacturing.

- The vehicle was successfully energized with a substitute battery, revealing the odometer reading and confirming the absence of a significant short circuit in the remaining vehicle harness.

- Indications of interventions on the battery connections and nearby harness were found, potentially related to installing and removing a tracking device or alarm.

- One threaded tie rod holding the battery bracket was incorrectly anchored, positioning the bracket too close to the positive pole connection.

- A short circuit between this metallic bracket and the positive pole connection could have ignited the fire.

- No similar incidents were found on an Internet search.

- Surveys of other vehicles from the same manufacturing period revealed the correct assembly of the battery brackets and their tie rods.

15.2.4.
Main Photos

As depicted in **Figure 15.1**, the engine hood underwent deformation during the fire suppression efforts. The black paint exhibited only signs of damage attributed to excessive temperatures within the engine compartment, with no discernible patterns indicative of arson.

Figure 15.1 Front view.

Courtesy of Erbis Llobet Biscarri.

Figure 15.2 illustrates that the cabin remained undamaged. The firewall and windshield effectively kept the cabin temperature low during the firefighting activities.

In **Figure 15.3**, the fire consumed a significant portion of combustible materials in the upper region of the engine compartment, including air ducts, air filter, fuel rail, various plastic parts, hoses, and harnesses. The lower area of the engine compartment showed minor damages and no signs of exposure to external flames.

Figure 15.2 Cabin interior.

Courtesy of Erbis Llobet Biscarri.

Figure 15.3 Engine compartment.

Courtesy of Erbis Llobet Biscarri.

Figure 15.4 showcases the remnants of the battery and emphasizes the close proximity between the positive pole connection and the edge of the metallic bracket above the battery.

In contrast, vehicles in standard condition typically exhibit a considerable clearance between the positive pole connection and this bracket.

Figure 15.4 Battery remnants and short circuit region.

Courtesy of Erbis Llobet Biscarri.

In **Figure 15.5**, the front threaded tie rod fragments, constructed from steel, are prominently featured. This indicates that the tie rod was subjected to a substantial electric current, which caused it to melt into pieces during the short circuit.

In **Figure 15.6**, captured post-battery removal, the incorrect positioning of the rear tie rod is emphasized in red. Additionally, a cut plastic clamp, highlighted in yellow, is observed. This clamp initially secured a horizontal cable to the battery support, and further evidence (not depicted here) suggests that this cable underwent handling and modifications.

Figure 15.5 Melted pieces of the front tie rod.

Courtesy of Erbis Llobet Biscarri.

Figure 15.6 Incorrect position of the rear tie rod and cut plastic clamp.

Courtesy of Erbis Llobet Biscarri.

15.2.5.
Hypotheses Testing

Each category group underwent a systematic analysis, considering the gathered evidence, effectively invalidating specific hypotheses. Additionally, this process allowed for the refinement of the root cause. The testing results are outlined in the following paragraphs.

15.2.5.1.
Starting Point

The location hypothesis S2 (external to the vehicle) was dismissed due to the absence of accelerant patterns on the hood and minimal damages on the lower region of the engine. Consequently, hypothesis S1 (inside the engine compartment) was upheld as the likely starting point of the fire, aligning with the owner's account and the manufacturer's report.

15.2.5.2.
Cause

The absence of recalls, the lack of similar events reported online, and the distinct positioning of battery attachments in regular vehicles led to the dismissal of hypotheses C1 (design error) and C2 (manufacturing error). Hypothesis C4 (deliberate) was also discarded due to the absence of accelerant signs and supporting evidence.

The progression of the analyses pointed to C3 (service error) as the likely cause, refined as follows:

The fire originated in the engine compartment due to a short circuit between the battery bracket and its positive pole connection. Post-manufacturing services on the battery and a nearby harness unintentionally left the rear threaded tie rod of the battery bracket anchored in an incorrect position, bringing the bracket closer to the pole connection. As the vehicle moved and parts underwent thermal expansion and contraction, the paint on the metallic bracket eroded, establishing a short circuit path from the positive pole of the battery through the bracket and the front tie rod to the vehicle chassis.

15.2.5.3.
Intent

With Hypothesis C4 (deliberate) eliminated, I1 (intentional) was also dismissed. Consequently, I2 (accidental) was upheld as the intent behind the incident.

15.2.6.
Conclusion

15.2.6.1.
Starting Point

The ignition originated in the engine compartment, stemming from a short circuit between the connection of the positive pole of the battery and the metallic bracket securing it in place.

15.2.6.2.
Cause

Post-manufacturing, inadvertent service activities involving the battery and nearby wiring harness led to the rear threaded tie rod of the battery bracket being incorrectly anchored, positioning the bracket too close to the pole connection. As the vehicle underwent movements and experienced thermal expansion and contraction of its components, the paint on the metallic bracket wore away, resulting in a short circuit that bridged from the battery's positive pole, through the bracket, and via the front tie rod to the vehicle chassis.

15.2.6.3.
Intent

The incident was deemed accidental.

15.3.
Case B: Engine Compartment Fire

Six years after being acquired by its initial owner, a subcompact hatchback car unexpectedly caught fire during a short trip. This incident unfolded the day after the vehicle received engine service. In reaction to the fire, the owner initiated legal action against the auto repair shop, attributing the blaze to the recent repair. The subsequent analysis, conducted three years after the event, sought to uncover the fundamental cause of the fire, aiming to furnish crucial evidence for a judicial resolution.

15.3.1.
Information Available before Vehicle Analysis

- Flex-fuel (gasoline/ethanol) ICEV, passenger car.

- Approximately 70,000 km and six years of use.

- The vehicle was uninsured at the time of the incident.

- The owner exhibited regular records of vehicle maintenance.

- The day before the event, the vehicle underwent a service involving the cleaning of fuel injectors to address the irregular engine power and unstable idling.

- Following the service, the vehicle was promptly taken from the repair shop to the owner's residence, less than 5 min away.

- The fire ignited during the subsequent trip, approximately 20 min into the drive, right in front of a police station.

- A bystander captured a brief video when the fire was limited to the engine compartment.

- A collision had occurred five years before the fire incident.

15.3.2.
Preliminary Hypotheses

The early-recorded video unequivocally captured the origin of the fire in the engine compartment during the initial stages of the incident. Consequently, there was no formal necessity to speculate on the ignition point. Given that the vehicle was in motion and the fire was observed in front of a police station, no hypotheses were generated concerning intent, and the fire was deemed accidental. Consequently, only the root cause was hypothesized. Moreover, the involved party in the legal proceedings had already posited specific hypotheses. By extrapolating from the available information, typical hypotheses were incorporated, resulting in the subsequent list:

- C1: Fuel leakage in recently serviced components.
- C2: Inadequate maintenance.
- C3: Damage to parts from a prior collision.
- C4: Electrical cause/short circuit.

15.3.3.
Key Findings during Vehicle Analysis and Owner's Interview

- The prior collision, which affected the left front corner of the vehicle, triggered the deployment of frontal airbags. Prompt repairs were conducted, and subsequent behavior exhibited no abnormalities for approximately five years until the fire occurred.

- The vehicle was devoid of any installed accessories.

- Approximately one year before the incident, the engine oil and filter were replaced.

- The fuel injectors were cleaned using an ultrasonic bath with an appropriate detergent fluid. No other components were replaced or cleaned.

- Given the vehicle's age, replacing the sealing O-rings of the fuel injectors and rail would have been prudent, even if their visual condition was satisfactory.

- No other mechanical or electrical failures were observed.

- A nearby vehicle alerted the driver to flames originating beneath the engine. The driver halted the car, exited the cabin, and then detected a strong gasoline odor. He attempted to put out the flames with a portable fire extinguisher. Unfortunately, he could not open the engine compartment and was unsuccessful.

- The fire progressed from the engine compartment to the cabin.

- The flames were ultimately extinguished by firefighters using water.

- The vehicle's rear section sustained less damage, with the owner selling the rear and spare tires separately.

- There were no damages to the engine block walls and carter.

- The fuel tank lid remained in good condition and securely in place.

- The plastic fuel tank exhibited no significant damage.

- The vehicle spent some time in suboptimal storage conditions, which led to theft of parts, vandalism, and weather-related degradation.

- The absence of several components and the time-related damage to several parts (e.g., engine harness) impaired the vehicle analysis.

- No recalls related to the issue were identified, and no comparable cases were reported on the Internet.

15.3.4.
Main Photos

Figures 15.7 and **15.8** show, respectively, the front and rear views of the vehicle on the start of the analysis. The vehicle had been severely rusted by exposure to the elements, and some parts were removed or stolen.

The engine compartment, seen in more approximation in **Figure 15.9**, had severe damages, and some of the parts and harnesses were consumed by the flames, lost during the precarious storage, or stolen.

The alternator, oil filter, and starter motor remained almost intact in the scene, as visible in **Figure 15.10**. This reduces the probability of a violent failure of one of these parts having ignited the fire.

Figure 15.11 highlights two deformed areas on the bumper's metallic core. These are consistent with the collision that occurred several years ago. However, it did not significantly affect the frame rails and crash boxes, resulting in relatively minor damage within the engine compartment.

Figure 15.7 Front view.

Courtesy of Erbis Llobet Biscarri.

Figure 15.8 Rear view.

Courtesy of Erbis Llobet Biscarri.

Figure 15.9 Engine compartment.

Courtesy of Erbis Llobet Biscarri.

Figure 15.10 Alternator, oil filter, and starter still in place.

Courtesy of Erbis Llobet Biscarri.

Figure 15.11 Front bumper deformations.

Courtesy of Erbis Llobet Biscarri.

Figure 15.12 captures the plastic tank lid and cap, found in place and good condition. Also, the plastic fuel tank did not exhibit signs of its rupture by flames, indicating that a tank leakage did not cause the fire.

Figure 15.12 Fuel tank lid in good condition.

Courtesy of Erbis Llobet Biscarri.

15.3.5.
Hypothesis Testing

Considering the significant time gap between the occurrence of the fire and the subsequent analysis, coupled with the precarious condition in which the vehicle was stored and the available evidence, none of the root cause hypotheses could be definitively ruled out with a high level of confidence. In essence, all hypotheses remained within the realm of possibility. In light of these circumstances, the most pragmatic approach was to assess the relative likelihood of each hypothesis, resulting in the following evaluations:

- Hypothesis C1 (fuel leakage on the fuel injectors or rail) emerged as the most probable cause. This conclusion was drawn based on various factors, including the recent service involving these components, the reuse of the old O-rings (increasing the likelihood of rupture or insufficient fuel sealing), the short period between the service and the fire incident, the ignition of the fire when the engine was at an elevated temperature, the observation of flames before the driver detected any abnormalities, and the distinct smell of gasoline on the driver's exit from the cabin.

- Hypothesis C2 (inadequate maintenance) was considered a plausible but less likely scenario, considering the vehicle's regular maintenance records and the considerable time interval since the last maintenance.

- Hypothesis C3 (damage to engine parts from the prior collision) was also deemed plausible but less probable, given the relatively minor damage to the engine compartment and the continued use of the vehicle for approximately five years post-repair.

- Hypothesis C4 (electrical cause/short circuit) was considered a potential but less likely factor. This assessment was based on the absence of detected anomalies before the emergence of flames and the lack of clear indications of internal damage to the alternator and starter motor remnants.

15.3.6.
Conclusion

In this case, the primary focus was to identify the root cause of the fire, which started in the engine compartment. The fire was considered accidental since it began in front of a police station while the vehicle was in motion.

The most likely hypothesis was fuel leakage on the O-rings of the fuel injectors or of the fuel rail since there was a recent service involving these components, the reuse of the old O-rings, the short period between the service and the fire incident, the ignition of the fire when the engine was at an elevated temperature, the observation of flames before the driver detected any abnormalities, and the distinct smell of gasoline on the driver's exit from the cabin.

The following causes were retained as possible but unlikely: inadequate maintenance (of other parts or systems), damage to engine parts from a collision five years before, and an electrical cause/short circuit.

It is noteworthy that despite exhaustive efforts, the typical objective of pinpointing the exact root cause was not achieved. Nevertheless, this case is included in this book as an illustrative example of handling scenarios where available evidence significantly hampers analysis outcomes.

15.4.
Case C: Truck Tractor Fire

A fire erupted behind the cabin of a truck tractor operating within a sizable fleet, transporting a load of garbage in the attached semitrailer. This incident occurred just minutes before the vehicle reached its destination, a landfill. A few months later, the fleet manager hired an investigation, with the primary objective of identifying the root cause of the fire and the secondary goal of preventing similar incidents for other vehicles within the fleet.

15.4.1.
Information Available before Vehicle Analysis

- The truck tractor had logged approximately 220,000 km over 2.5 years of operation.

- Before its inclusion in the fleet, a hydraulic system and reservoir were installed to enable the vertical movement of the semitrailer's front end.

- Toward the conclusion of an hour-long journey, an electrical malfunction occurred, shutting down most of the electric system and the diesel engine. Only a limited number of warning lights remained active.

- The driver's attempt to restart the engine using the ignition key proved futile. Subsequently, he exited the cabin and endeavored to reset the master electric switch without successfully restoring regular operation. This switch is located behind the left side of the cabin.

- A phone call to headquarters instructed him to await assistance. Roughly 3 min later, a nearby driver alerted him to the presence of visible flames.

- A cellular phone video documented the origin of the fire behind the right side of the cabin.

- Despite the driver's efforts with two portable fire extinguishers, the flames could not be extinguished.

- The fire progressed to the engine compartment, engulfing the entire cabin.

- No mechanical anomalies were observed before the onset of the fire.

- There were no pending recalls or similar incidents reported on the Internet.

- No collisions were recorded.

15.4.2.
Preliminary Hypotheses

A bystander's early-recorded video vividly captured the inception of the fire, definitively pinpointing its origin behind the right side of the cabin. Consequently, there was no imperative to conjecture about the ignition point. Given the vehicle's proximity to its intended destination at the time of the incident and the credibility of the driver's report, no suppositions were made regarding intent; the fire was attributed to an accidental cause.

In light of these circumstances, the focus shifted solely to identifying the root cause, resulting in the formulation of the following hypotheses:

- C1: Fuel leakage.
- C2: Electric cause/short circuit.
- C3: Debris in contact with hot engine or exhaust parts.
- C4: Service error.
- C5: Inadequate installation of accessories.
- C6: Proximity to flames or suspended embers.

15.4.3.
Key Findings during Vehicle Analysis and Driver's Interview

- Cabin and engine compartment severely damaged by the flames.

- The hydraulic system (which tilts the semitrailer) was barely damaged. There were no signs of fluid leakage.

- The alternator wiring harness was interrupted next to the metallic mount that secures the automatic transmission to the chassis rail. The remains of the main wire exhibited copper beads, and corresponding short circuit marks were identified on the metallic mount next to the wire.

- The alternator wire segment between this transmission mount and the battery was disintegrated, even though this trail was not exposed to flames from diesel combustion.

- Flames severely consumed the diesel fuel filter, whose feed and return lines crossed over the same transmission mount.

- The driver provided additional details, dismissing the likelihood of the landfill flames (intentionally created to eliminate naturally occurring combustible gases resulting from debris decomposition) or any potentially

suspended embers being the cause. This determination was based on the burners being situated far from the scene and the burning process not generating any suspended embers. Also, no wildfires or plantation fires occurred along the truck route.

- Services and repairs made before the incident had no relationship with the starting point of the fire.

- The semitrailer had minor damages and was already repaired when the truck tractor was analyzed. The semitrailer had returned to everyday use without the occurrence of new incidents.

15.4.4.
Main Photos

The cabin and engine compartment were severely burned, as seen in **Figures 15.13**, **15.14**, and **15.15**. The front tires, consumed by the fire,

Figure 15.13 Front view.

Courtesy of Erbis Llobet Biscarri.

Figure 15.14 Rear view.

Figure 15.15 Engine compartment.

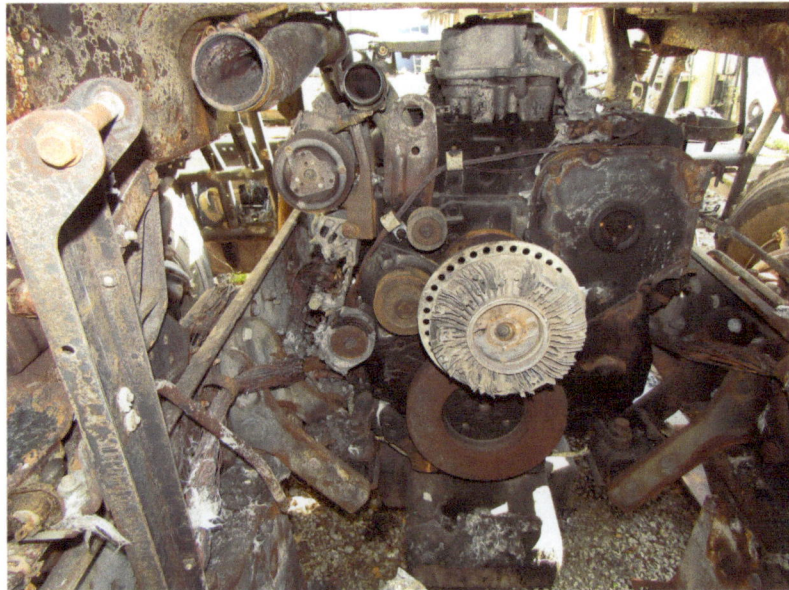

were replaced to allow the transportation of the vehicle remnants.

Figure 15.16 illustrates the remnants of the diesel filter: only the lower metallic body is discernible. This filter is in the region where the initial flames were spotted. **Figure 15.17** exhibits a filter in good condition from a similar vehicle, with the glass bowl on its upper area, allowing the vision of the fuel. The same figure highlights in red the diesel feed and return lines that cross the right rail through an oval opening.

Figure 15.18 highlights in yellow the remains of the alternator wire: only the segment between the alternator and the transmission's mount was found. The mount is emphasized in blue. The wire section starting around this mount, toward the truck's rear end (simulated by a red line), was disintegrated. This mount is also close to the region where the first flames were noticed.

Figure 15.19 shows a magnified view of the same region, while **Figure 15.20**, in greater magnification, points to copper beads in red and highlights craters on the mount edges in yellow.

Figure 15.16 Diesel filter remains.

Courtesy of Erbis Llobet Biscarri.

Figure 15.17 Diesel filter lines of a reference vehicle.

Courtesy of Erbis Llobet Biscarri.

Figure 15.18 Alternator wire over the transmission's mount.

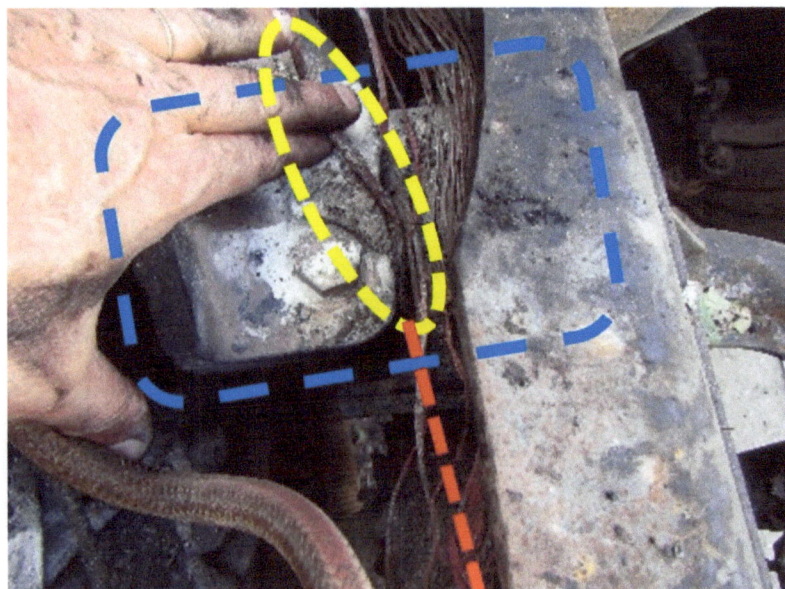

Courtesy of Erbis Llobet Biscarri.

Figure 15.19 Short circuit region.

Courtesy of Erbis Llobet Biscarri.

Figure 15.20 Copper beads on the wire and craters on the edges of the mount.

Courtesy of Erbis Llobet Biscarri.

These signs indicate a significant short circuit between the alternator wire and the chassis through the edges of the steel mount.

15.4.5.
Hypothesis Testing

Hypothesis C4 (service error) was rejected since the service records did not indicate any intervention related to the region where the first flames were observed.

Hypothesis C5 (inadequate installation of accessories) was eliminated since the only accessory (hydraulic lift system) was installed several years before, and its components did not reveal fluid leakages.

Hypothesis C6 (proximity to flames or suspended embers) was discarded since the landfill gas burn towers were far away from the scene, and no wildfires or plantation fires occurred that day.

Hypothesis C2 (short circuit) was considered the most likely cause since an electric malfunction occurred a few minutes before flames were noticed. This hypothesis was refined to a short circuit between the alternator wire and the right mount of the transmission. However, the possibility of this short circuit being caused as secondary damage could not be ruled out.

Hypothesis C1 (fuel leakage) was also considered possible, given the proximity of the diesel filter to the starting point of the fire. Yet it was less likely since the driver did not notice flames when the electric malfunction occurred or when he was deactivating the master electric switch behind the cabin.

Hypothesis C3 (debris in contact with hot engine or exhaust parts) was also considered possible since the semitrailer was carrying a load of debris and the starting point of the fire was next to the NOx exhaust reduction system (Arla 32™), which operates at high temperature. Since the driver did not visualize flames when the electric malfunction manifested nor when he was manipulating the master electric switch, this hypothesis was also considered less likely.

15.4.6.
Conclusion

In this investigation, the primary focus was on the root cause of the fire. The leading hypothesis retained was a short circuit between the alternator wire and the vehicle's chassis, triggered by friction damage of the wire insulator against the edges of the metallic mount of the transmission.

Another possible cause retained was a diesel leak from the filter feed and return lines rubbing against the edges of an oval opening in the right rail. However, this scenario was deemed less probable. Similarly, the hypothesis involving debris falling onto the exhaust pipes adjacent to the NOx reduction system was also considered possible and less likely.

A proactive recommendation was issued to inspect all remaining vehicles in the fleet. This inspection should ensure the proper securing of the alternator harness, using plastic clamps, to keep it at a safe distance from the moving edges of the transmission right mount. Additionally, during the same inspection, the implementation of a gasket on the edges of the oval opening in the rails, where the diesel filter lines intersect, was planned.

15.5.
Case D: Semitrailer Tire Fire

A fire broke out in the rear tires of a semitrailer that was part of a large fleet while transporting a load of sand. The maintenance manager commissioned an investigation a few months later, primarily focusing on pinpointing the root cause of the fire. The secondary intent was to assess the risk of similar incidents for other vehicles within the same fleet.

15.5.1.
Information Available before Vehicle Analysis

- The tractor truck and semitrailer had approximately 65,000 km and six months of use.

- The driver heard an explosion and spotted flames through the rearview mirror while driving the truck along a dirt road for about 10 min.

- A cellular phone video documented the fire starting on the last right-side tires.

- The driver unsuccessfully attempted to extinguish the flames: first with a portable fire extinguisher, then with sand and dirt.

- The fire subsequently spread to the left-side tires.

- A nearby water truck was dispatched and successfully extinguished the fire.

- No mechanical or electrical anomalies were observed before the fire.

- The manufacturers partially analyzed the semitrailer and its wheels, but their reports did not clearly indicate the root cause.

- No pending recalls or similar incidents were reported on the Internet.

- No collisions occurred.

- Similar vehicles in the same fleet exhibited premature wear-out of brake shoes.

15.5.2.
Preliminary Hypotheses

A video recorded early in the incident by the passenger (a coworker) documented that the fire originated from the rear inner tire on the right side. Consequently, there was no formal need to create hypotheses for the starting location. Considering the vehicle was in motion when the fire occurred, and the workers' reports were deemed reliable, no hypotheses were formulated regarding intent; the fire was considered accidental.

Therefore, only the root cause was hypothesized, as follows:

- C1: Electric cause/short circuit.

- C2: Wheel overheating due to insufficient lubrication.

- C3: Wheel overheating due to a failure in its braking system.

- C4: Brake/wheel lockup leading to tire overheating (due to its friction against the soil).

- C5: External to the vehicle/nearby flames.

15.5.3.
Key Findings during Vehicle Analysis

- Tire damage consistent with the explosion sound reported by the driver.

- More pronounced wear on brake shoes in the wheels of the rearmost axle.

- Overheated shoe on the rear right wheel.

- Oxidation and the absence of grease residues on the inner part of the same wheel drum indicated internal overheating.

- Ball bearings in good condition with no signs of abnormal wear or blocking.

- Parking and service brake control module in good condition.

- Wiring harness in good condition, away from the burnt area, showing no signs of short circuits.

- No flammable fluids in the burnt area.

- S-cam thrust washer with a flat shape and a large outline.

15.5.4.
Main Photos

Figure 15.21 displays the left-side view of the semitrailer as observed at the onset of the vehicle analysis. The damages on this side resulted from the delayed fire suppression efforts. Also, some remnants of tires and wheel parts were removed during the manufacturers' analysis.

Figure 15.22 depicts the rear right corner, extensively and more severely burned, consistent with the ignition of the fire in this region. Some wheel parts from this side were removed and transported to the maintenance office of the fleet.

Figure 15.23 illustrates one of the brake shoes stored in the office. In addition to significant wear, the cracks and missing pieces indicate overheating.

Figure 15.24 exhibits the S-cam brake assembly of a paradigm vehicle of a similar manufacturing period. In this photo, the brake shoes are still in place.

Figure 15.21 Left-side view.

Courtesy of Erbis Llobet Biscarri.

Figure 15.22 Rear view.

Courtesy of Erbis Llobet Biscarri.

Figure 15.23 Overheated brake shoe.

Courtesy of Erbis Llobet Biscarri.

Figure 15.24 S-cam brake assembly of a paradigm trailer.

Courtesy of Erbis Llobet Biscarri.

In **Figure 15.25**, the brake shoes were removed, and the salient edge of the S-cam thrust washer is highlighted. According to the wheel design documents, this washer should have bent edges to avoid mechanical interference with the shoes if the washer rotates. However, this was not the case in the vehicle analyzed, and the washers were flat. Manually rotating the washer confirmed the possibility of its edge holding the brake shoes in the brake position. If this occurs while the vehicle is in motion, the brake shoes will be continuously pressed against the inner face of the drum, rapidly increasing its temperature. After a few minutes, the drum temperature becomes sufficient to ignite the tire attached to it, initiating a vehicle fire.

Figure 15.25 Edge of the S-cam thrust washer.

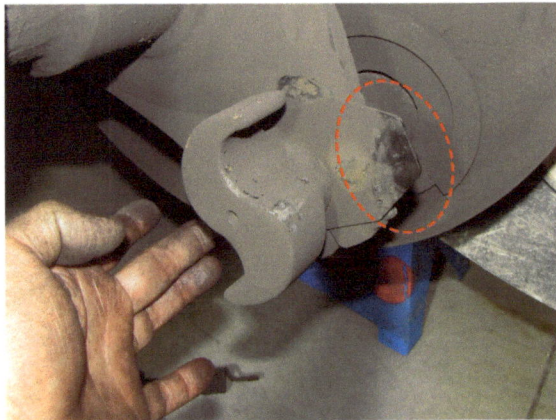

Courtesy of Erbis Llobet Biscarri.

15.5.5.
Hypothesis Testing
Hypothesis C1 (electric cause/short circuit) was rejected because no harnesses were deemed close enough to the wheels, therefore uncapable of overheating the tires. Additionally, there were no signs of short circuits or overload of the wires between the axles and the trailer connection point.

Hypothesis C2 (wheel overheating due to insufficient lubrication) was eliminated since there were no signs of excessive wear or lockup of the ball bearings of the wheels, situations that would have occurred in case of a lack of lubrication.

Hypothesis C4 (brake lockup leading to tire overheating) was discarded due to the overheating of the brake shoes since a locked-up wheel does not overheat its brake shoes. Also, the parking and safety brake module were in good condition.

Hypothesis C5 (external to the vehicle/nearby flames) was excluded since the vehicle did not encounter any wild or plantation fire on the relevant date.

Hypothesis C3 (wheel overheating due to a failure in its braking system) was retained as the root cause and refined as follows: the brake shoes overheated due to an incorrect geometry of the S-cam thrust washer, allowing mechanical interference of its edge with the brake shoes. Consequently, after some braking events, the shoes remained pressed against the drum, rapidly elevating its temperature since the vehicle continued to move. After some minutes in this condition, the drum temperature was sufficient to ignite the tire.

15.5.6.
Conclusion
In this case, the investigation initially focused solely on identifying the root cause. It was determined that the primary factor was the incorrect shape of the thrust washer between the S-cam and the wheel plate. During brake usage, this washer rotated, and its edge interfered with the brake shoe, causing it to remain pressed against the drum after the brake pedal was released. A recommendation was promptly issued to inspect the remaining vehicles in the fleet urgently and replace flat thrust washers with those featuring a shorter outline (i.e., with bent edges).

15.6.
Case E: BEV Fire after Running over Debris

BEV fires remain relatively rare, despite the extensive media coverage and attention. Consequently, opportunities to thoroughly analyze, document, and share findings related to BEV fires are limited. In September 2023, a BEV encountered metallic debris on an Australian highway, damaging its battery pack and triggering a thermal runaway. Remarkably, the occupants emerged from the vehicle unharmed, and the prompt intervention of a fire brigade curbed the flames, mitigating the complete damage to the car. The manufacturer, who was also the owner of the BEV in this instance, generously provided the remnants for a meticulous analysis by EV Fire Safe, a highly regarded Australian organization. This organization is renowned for maintaining an extensive global database documenting incidents of xEV fires and offering valuable training and consultancy services within this domain.

Here are some of the details of the EV Fire Safe analysis, which was conducted with the manufacturer's support [15.7].

15.6.1.
Information Available before Vehicle Analysis

- The accident occurred in September 2023 on an Australian highway.
- The car, traveling at a speed of 111 km/h, ran over a substantial steel part that had fallen from a trailer carrying loose components.

- This collision triggered a thermal runaway within the battery, leading to the eventual ignition of its gases. It took several minutes for the resulting flames to engulf the car.
- Fortunately, no one was injured.
- The battery pack consisted of 4416 cylindrical lithium-ion cells organized into four longitudinal modules beneath the vehicle cabin.
- The baseplate of the battery pack is made of aluminum, with a melting point around 660°C. In case of a battery fire, the plate was designed to melt, allowing the cells to "drop out," facilitating the cooling of the battery pack. This design was also meant to reduce the spread of a thermal runaway to adjacent cells.

15.6.2.
Key Findings during Vehicle Analysis

- The responsible debris, an 18-kg truck tail shaft, was recovered from the accident scene.
- The collision caused significant damage, tearing the lower aluminum plate and impacting multiple battery cells. This resulted in short circuits and triggered a thermal runaway.
- The cabin's interior sustained relatively minor damage, and remarkably, the cabin floor remained intact.
- Flames started penetrating the cabin through the HVAC opening on the firewall.
- Despite the severity of the incident, the airbags were not deployed, indicating that there was no significant impact or substantial deceleration in the vehicle's cabin. The lower region of the vehicle sustained severe damage, but the lack of airbag deployment confirms that the cabin's

safety thresholds were not met. On the other hand, the pyrofuse was engaged (as expected), promptly disconnecting the high-voltage system from the compromised battery pack.

- The vehicle analysis, conducted two months post-incident, revealed that several battery cells still retained a charge (i.e., contained "stranded energy"). Therefore, a reignition was possible.

- Precautions taken by the analysis team are listed as follows:

  - Utilizing appropriate PPE such as full-body suits, filtered masks, safety glasses, high-voltage gloves, boots, and wipes.

- Touching vehicle components with only one hand to minimize the risk of electrically bridging any energized part to the chassis.

- Regularly mapping the temperature of the remaining battery cells.

- Handling each battery cell carefully due to the potential risk of thermal runaway and explosion.

15.6.3.
Main Photos

Figure 15.26, a frontal perspective, reveals the most severely damaged area. After the collision, firefighting, and relocation of the wreck, the front right wheel and its suspension disengaged from the vehicle body.

Figure 15.26 Front view.

Courtesy of Emma Sutcliffe, Director, evfiresafe.com.

Figure **15.27** provides an insight into the cabin's interior, showcasing heightened damage to the instrument panel and windshield. Notably, a plastic water bottle in the center console remained intact.

A bottom view in Figure **15.28** captures the car suspended in a suitable elevator. The lower aluminum plate exhibits perforations in numerous locations, which facilitated the drop of several cells from the battery pack.

Figure **15.27** Cabin interior.

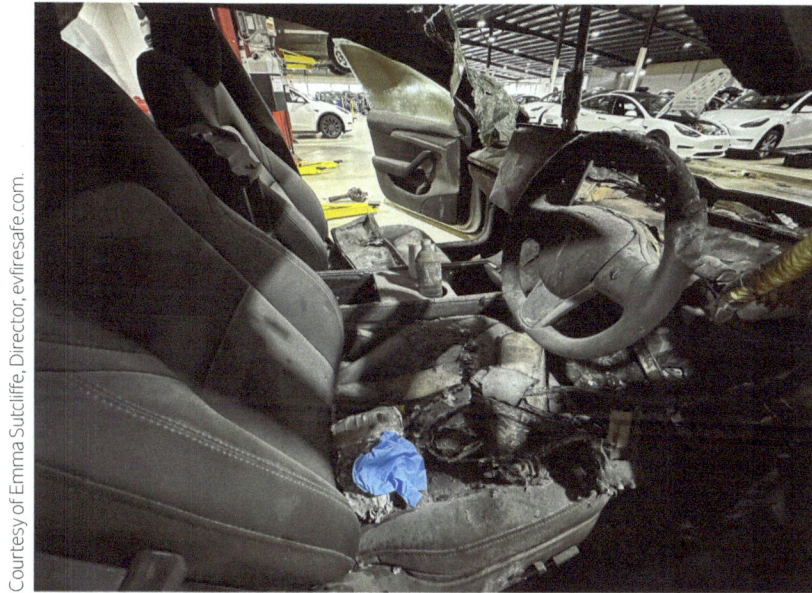

Courtesy of Emma Sutcliffe, Director, evfiresafe.com.

Figure **15.28** Damaged and melted aluminum plate.

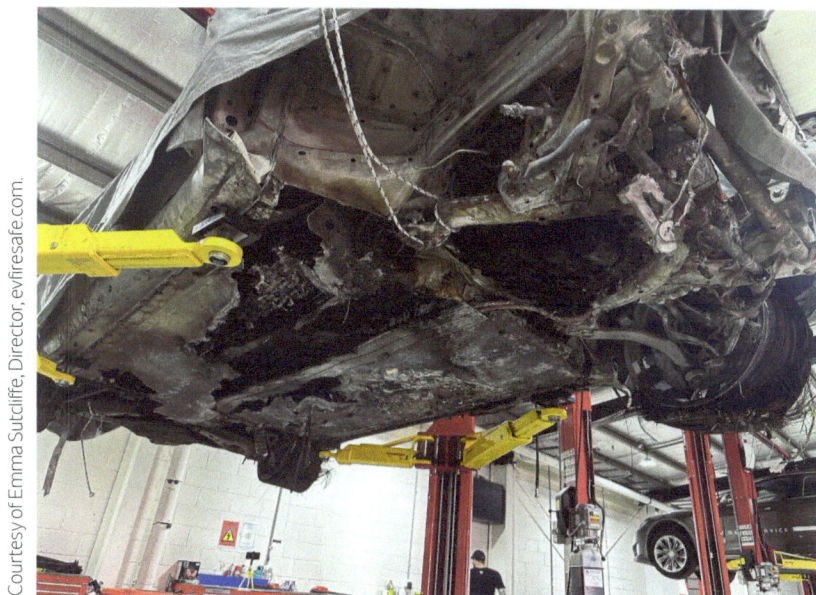

Courtesy of Emma Sutcliffe, Director, evfiresafe.com.

Figure 15.29 illustrates the relative trajectory of the metallic debris, which carved through one of the battery modules from its front to its rear end.

Figure 15.29 Relative trajectory of the debris.

Courtesy of Emma Sutcliffe, Director, evfiresafe.com.

Finally, **Figure 15.30** depicts the large metallic road debris held near its final position, conclusively confirming its role in causing significant mechanical destruction during the collision.

15.6.4. Conclusion

Diverging from the approach taken in the previous case analyses, the focus of this case did not involve pinpointing the origin of the fire, determining its root cause, or deciphering any malicious intent, as all these aspects were already established. However, it is anticipated that the insights and details presented here will benefit many readers.

Figure 15.30 Final position of the debris.

Courtesy of Emma Sutcliffe, Director, evfiresafe.com.

References

[15.1]. Barnett, G., *Vehicle Battery Fires: Why They Happen and How They Happen*, SAE International R-443 Book (Warrendale: SAE International, 2017), ISBN:978-0-7680-8143-5.

[15.2]. Parrott, K. and Stahl, D., "Case Studies of Parking Brake Fires in Commercial Vehicles," SAE Technical Paper 2013-01-0207 (2013), doi:https://doi.org/10.4271/2013-01-0207.

[15.3]. Scheibe, R., Shields, L., and Angelos, T., "Field Investigation of Motor Vehicle Collision-Fires," SAE Technical Paper 1999-01-0088 (1999), doi:https://doi.org/10.4271/1999-01-0088.

[15.4]. Rae-Tech Fire Investigations Ltd., "Burned Vehicle Examination VTS 04 1," accessed March 2024, https://youtu.be/WZ5XaZF_L0M.

[15.5]. Rae-Tech Fire Investigations Ltd., "Burned Vehicle Examination # 2VTS 05 1," accessed March 2024, https://youtu.be/pchqAHObmYo.

[15.6]. Rae-Tech Fire Investigations Ltd., "Vehicle Fire Investigation #3 VTS 10 1," accessed March 2024, https://www.youtube.com/watch?v=w81NLUSCmcQ.

[15.7]. NSW Association of Fire Investigators (NSW AFI) and EV FireSafe, "February 2024 Education Night—Case Study: Tesla Model 3 Incident, NSW AFI, EV Fire Safe," accessed March 2024, https://www.youtube.com/watch?v=jJkOk2q5lEw&t=3479s.

Bibliography

Barnett, G., *Vehicle Battery Fires: Why They Happen and How They Happen*, SAE International Book R-443 (Warrendale: SAE International, 2017), ISBN:978-0-7680-8143-5.

Robert Bosch GmbH, *Bosch Automotive Handbook*, 10th ed. (Plochingen, Germany: Robert Bosch GmbH, 2018), ISBN:978-0-7680-9567-8, distributed by SAE International.

Brach, M., *SAE International's Dictionary of Vehicle Accident Reconstruction and Automotive Safety*, SAE International Book R-556 (Warrendale: SAE International, 2023), ISBN:978-1-4686-0594-5.

Brach, M., Brach, R., and Mason, J., *Vehicle Accident Analysis and Reconstruction Methods*, 3rd ed., SAE International Book R-516 (Warrendale: SAE International, 2022), ISBN:978-1-4686-0419-1.

Chang, C.-H., Gorin, C., Zhu, B., Beaucarne, G. et al., "Lithium-Ion Battery Thermal Event and Protection: A Review," *SAE Int. J. Elec. Veh.* 13, no. 3 (2024): 1-41, doi:https://doi.org/10.4271/14-13-03-0019.

De Santis, T., Adams, C., Molnar, L., Washington, J. et al., "Motor Vehicle Fire Investigation," SAE Technical Paper 2008-01-0555 (2008), doi:https://doi.org/10.4271/2008-01-0555.

Gann, R. and Friedman, R., *Principles of Fire Behavior and Combustion* (Burlington, MA: Jones & Bartlett Learning, 2013), ISBN:1284056104, 9781284056105.

Icove, D. and Haynes, G., *Kirk's Fire Investigation*, 8th ed. (New York: Pearson, 2018), ISBN 10:0-13-423792-7 and ISBN 13:978-0-13-423792-3.

Icove, D., Haynes, G., and De Han, J., *Forensic Fire Scene Reconstruction*, 3rd ed. (Boston: Pearson, 2012), ISBN-13:9780132605779.

International Electrotechnical Commission, "Amendment 5—International Electrotechnical Vocabulary (IEV)—Part 151: Electrical and Magnetic Devices," Standard IEC 60050-151:2001/AMD5:2021, 2021.

International Organization for Standardization, "Road Vehicles—Electrical Disturbances from Conduction and Coupling—Part 1: Vocabulary and General Considerations," Standard ISO 7637-1:2023(en), 2023.

Kershaw, J., *SAE International's Dictionary for Automotive Engineers*, SAE International Book R-523 (Warrendale: SAE International, 2023), ISBN:978-1-4686-0407-8.

National Fire Protection Association, "Guide for Fire and Explosion Investigations—NFPA 921," 2021.

National Fire Protection Association, "Standard for Professional Qualifications for Fire Investigator—NFPA 1033," 2022.

Noon, R., *Forensic Engineering Investigation*, 1st ed. (Boca Raton: CRC Press, 2000), doi:https://doi.org/10.1201/9781420041415.

SAE International Ground Vehicle Standard, "Hybrid Electric Vehicle (HEV) and Electric Vehicle (EV) Terminology," SAE Standard J1715, J1715_202209, September 2022.

SAE International Ground Vehicle Standard, "Battery Terminology," SAE Standard J1715/2, J1715/2_202108, August 2021.

SAE International Ground Vehicle Standard, "Guidelines for Electric Vehicle Safety," SAE Standard J2344, J2344_202010, October 2020a.

SAE International Ground Vehicle Standard, "Hybrid and EV First and Second Responder Recommended Practice," SAE Standard J2990, J2990_201907, July 2019.

SAE International Ground Vehicle Standard, "Hybrid and Electric Vehicle Safety Systems Information Report," SAE Standard J2990/2, J2990/2_202011, November 2020b.

SAE International Ground Vehicle Standard, "Best Practices for Storage of Lithium-Ion Batteries," SAE Standard J3235, March 2023, https://doi.org/10.4271/J3235_202303.

Scheibe, R., Shields, L., and Angelos, T., "Field Investigation of Motor Vehicle Collision-Fires," SAE Technical Paper 1999-01-0088 (1999), doi:https://doi.org/10.4271/1999-01-0088.

Struble, D. and Struble, J., *Automotive Accident Reconstruction: Practices and Principles* (Boca Raton: CRC Press, 2020), doi:https://doi.org/10.1201/9781003008972.

Sun, P., Bisschop, R., Niu, H., and Huang, X., "A Review of Battery Fires in Electric Vehicles," *Fire Technology* 56 (2020): 1361-1410, doi:https://doi.org/10.1007/s10694-019-00944-3.

Womack, J., Jones, D., and Roos, D., *The Machine That Changed the World* (London, UK: Simon & Schuster, 2007), ISBN:978-0743299794.

Acronym Glossary

ABS - (1) Antilock Brake System

(2) Acrylonitrile Butadiene Styrene (a polymer)

AC - Alternating Current

ACEA - European Automobile Manufacturers' Association

ADAS - Advanced Driver Assistance System

AFV - Alternative Fuel Vehicle

AGM - Absorbed Glass Mat (a type of lead-acid battery)

ANSI - American National Standards Institute

APU - Auxiliary Power Unit

ASTM - American Society for Testing and Materials

ATA - American Trucking Associations

BCM - Body Control Module

BCU - Battery Control Unit

BEV* - Battery Electric Vehicle

BFP - Bureau of Fire Protection

BLEVE - Boiling Liquid Expanding Vapor Explosion

BMS* - Battery Management System

BTS - Bureau of Transportation Statistics

BTU - British Thermal Unit

CAD - Computer-Aided Design

CAE - Computer-Aided Engineering

CAFI - Canadian Association of Fire Investigators

CAN - Controller Area Network

CCA - *Court Circuit Aggravé* (aggravated short-circuit)

CCS - Combined Charging System

CCTV - Closed-Circuit Television

CFD - Computational Fluid Dynamics

CFPA - Confederation of Fire Protection Associations Europe

CFR - Code of Federal Regulations

CNG - Compressed Natural Gas

CTIF - *Comité Technique International de Prévention et d'Extinction du* Feu (International Association of Fire and Rescue Services)

CV - Constant Velocity joint

DBI - Danish Institute of Fire and Security Technology

DC - Direct Current

DFMEA - Design Failure Modes and Effects Analysis

DoT or DOT - Department of Transportation

DPF - Diesel Particulate Filter

DTC - Diagnostic Trouble Code

DWPT - Dynamic Wireless Power Transfer

ECA - *Échauffement Aggravé* (aggravated overheating)

ECE - Economic Commission for Europe

ECM - Engine Control Module

ECU - Electronic Control Unit

EDR - Event Data Recorder

EFI - European Forest Institute

EGR - Exhaust Gas Recirculation

EMC - Electromagnetic Compatibility

EMI - Electromagnetic Interference

EOL - End of Life

EPS - (1) Electric Power Steering

(2) Electric Power System

ESD* - Electrostatic Discharge

EU - European Union

EV* - Electric Vehicle

FCEV or FCV* - Fuel Cell Electric Vehicle

FMEA - Failure Modes and Effects Analysis

FMVSS - Federal Motor Vehicle Safety Standard

GM - General Motors

HEV* - Hybrid Electric Vehicle

HMI - (1) Human–Machine Interaction
(2) Human–Machine Interface

HMIS - Hazardous Materials Information System

HPU - Hybrid Power Unit

HRR* - Heat Release Rate

HV - Hybrid Vehicle

HVAC - Heating, Ventilation, and Air Conditioning

HVIL - High-Voltage Interlock Loop

IAAI - International Association of Arson Investigators

IAFC - International Association of Fire Chiefs

IAFI - International Association of Fire Investigators

ICE - Internal Combustion Engine

ICEV* - Internal Combustion Engine Vehicle

ICV* - Internal Combustion (engine) Vehicle

IEC - International Electrotechnical Commission

IFSSC - International Fire Safety Standards Coalition

IP - Instrument Panel

IPD - Integrated Product Development

IPXX* - Ingress Protection classification

IR - Infrared Radiation

ISO - International Organization for Standardization

LCO - Lithium Cobalt Oxide

LED - Light Emitting Diode

LFL* - Lower Flammability Limit

LFP - Lithium Iron Phosphate

LIB* - Lithium-Ion Battery

LMO - Lithium Manganese Oxide

LNG - Liquified Natural Gas

LPG - Liquified Petroleum Gas

MHEV* - Micro Hybrid Electric Vehicle

MHV or mHEV* - Mild Hybrid Vehicle

MIL - Malfunction Indicator Lamp

MSD* - Manual Service Disconnect (of the high voltage)

MY - Model Year

NACS* - North American Charging Standard (PEV connection)

NAFI - National Association of Fire Investigators

NCA - Nickel Cobalt Aluminum

NEV* - New Energy Vehicle

NFPA - National Fire Protection Association

NFSTI - National Forensic Science Training Institute

NHTSA - National Highway Traffic Safety Administration

NIST - National Institute of Standards and Technology

NMC - Nickel Manganese Cobalt

NMH - Nickel-metal hydride

NOx - Nitrogen oxides

NTSB - National Transportation Safety Board

OBC* - On-Board Charger

OBD - On-Board Diagnostics

OEM - Original Equipment Manufacturer

OSHA - Occupational Safety and Health Administration

PA - Polyamide (a polymer)

PCB - Printed Circuit Board

PCM - Powertrain Control Module

PE - Polyethylene (a polymer)

PEV* - Plug-in Electric Vehicle

PFMEA - Process Failure Modes and Effects Analysis

PHEV* - Plug-in Hybrid Electric Vehicle

PHRR* - Peak Heat Release Rate

PP - Polypropylene (a polymer)

PPE - Personal Protective Equipment

PPTC - Polymeric Positive Temperature Coefficient device (also known as polyswitch)

PTFE - Polytetrafluoroethylene (a polymer)

PVC - Polyvinyl Chloride (a polymer)

R&D - Research and Development

RESS - Rechargeable Energy Storage System

RFI - Radio Frequency Interference

RISE - Research Institutes of Sweden

SCR - Selective Catalytic Reduction

SLA - Sealed Lead Acid (a type of lead–acid battery)

SoC or SOC - State of Charge (of the battery)

SoH or SOH - State of Health (of the battery)

SUV - Sport Utility Vehicle

TMC - Technology & Maintenance Council—American Trucking Organization

TQM - Total Quality Management

UFL* - Upper Flammability Limit

UK - United Kingdom

UL - Underwriters Laboratories

US - United States

USABC - United States Advanced Battery Consortium

USFA - US Fire Administration

UV - Ultraviolet Radiation

VIN - Vehicle Identification Number

VRLA - Valve Regulated Lead Acid (a type of lead–acid battery)

WCS - Wireless Charging System

WPT - Wireless Power Transfer

xEV* - "Any" Electrified Vehicle

ZEV* - Zero-Emission Vehicle

Index

About the Author

Erbis Llobet Biscarri

Erbis Llobet Biscarri boasts a distinguished four-decade career in technical and engineering domains, offering expertise in automotive and electronic design, quality issues

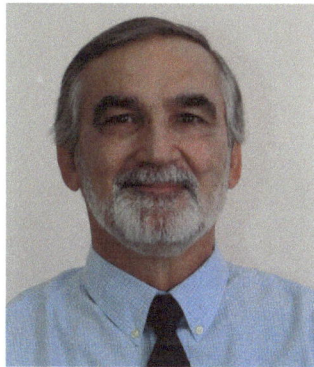

Courtesy of Erbis Llobet Biscarri.

resolution, electromagnetic compatibility, and fire prevention and analysis. Holding a master's degree in manufacturing systems engineering from the University of Michigan and degrees in industrial administration and electronics engineering from the University of São Paulo, this professional blends academic excellence with practical knowledge.

Contributing significantly to Ford group companies worldwide, he played pivotal roles in design and development at Ford Electronics and Visteon. Johnson Controls Automotive–Electronics saw his tenure as the engineering manager in South America, overseeing local R&D and contributing to successful product launches.

At PSA Peugeot Citroën in Brazil, leadership roles included responsibilities in electro-electronics architecture and oversight of EE program managers in Latin America. Simultaneously, serving as a technical specialist for hardware development, he played a vital role in a team dedicated to analyzing and preventing car fire incidents, showcasing the benefits of collective expertise and fostering collaborative learning.

Beyond corporate roles, the author contributed to SAE congresses in Brazil by holding a position on the technical congress committee and serving as a Porto Alegre section board director. The author actively engaged in SAE International congresses in Brazil and the United States and shared insights on instrument clusters, quality, and hybrid electric vehicle simulation.

In recent years, his journey gained new dimensions. He contributes as a forensic engineering expert in Brazil's judicial system and as an SAE instructor, focusing on automotive and agricultural machine fire theory, prevention, and analysis. These diverse professional activities underscored the necessity for the comprehensive book presented here.

www.ingramcontent.com/pod-product-compliance
Lightning Source LLC
Chambersburg PA
CBHW042058210326
41597CB00045B/35